Food Price Dynamics and Price Adjustment in the EU

Food Price Dynamics and Price Adjustment in the EU

Edited by
Steve McCorriston

OXFORD
UNIVERSITY PRESS

Great Clarendon Street, Oxford, OX2 6DP,
United Kingdom

Oxford University Press is a department of the University of Oxford.
It furthers the University's objective of excellence in research, scholarship,
and education by publishing worldwide. Oxford is a registered trade mark of
Oxford University Press in the UK and in certain other countries

© the various contributors 2015

The moral rights of the authors have been asserted

First Edition published in 2015

Impression: 1

All rights reserved. No part of this publication may be reproduced, stored in
a retrieval system, or transmitted, in any form or by any means, without the
prior permission in writing of Oxford University Press, or as expressly permitted
by law, by licence or under terms agreed with the appropriate reprographics
rights organization. Enquiries concerning reproduction outside the scope of the
above should be sent to the Rights Department, Oxford University Press, at the
address above

You must not circulate this work in any other form
and you must impose this same condition on any acquirer

Published in the United States of America by Oxford University Press
198 Madison Avenue, New York, NY 10016, United States of America

British Library Cataloguing in Publication Data
Data available

Library of Congress Control Number: 2015936103

ISBN 978–0–19–873239–6

Printed and bound by
CPI Group (UK) Ltd, Croydon, CR0 4YY

Links to third party websites are provided by Oxford in good faith and
for information only. Oxford disclaims any responsibility for the materials
contained in any third party website referenced in this work.

Contents

List of Figures vii
List of Tables ix
List of Contributors xi

1. Introduction 1
   Steve McCorriston

2. Food Inflation in the EU: Contrasting Experience and
   Recent Insights 20
   Tim Lloyd, Steve McCorriston, and Wyn Morgan

3. Overview of Price Transmission and Reasons for Different
   Adjustment Patterns across EU Member States 51
   Islam Hassouneh, Carsten Holst, Teresa Serra, Stephan von
   Cramon-Taubadel, and José M. Gil

4. Price Transmission in Food Chains: The Case of the Dairy Industry 65
   Céline Bonnet, Tifenn Corre, and Vincent Réquillart

5. Spatial and Temporal Retail Pricing on the German Beer Market 102
   Jens-Peter Loy and Thomas Glauben

6. The Use of Scanner Data for Measuring Food Inflation 122
   Elena Castellari, Daniele Moro, Silvia Platoni, and Paolo Sckokai

7. Price Transmission in Modern Agricultural Value Chains:
   Some Conceptual Issues 147
   Johan Swinnen and Anneleen Vandeplas

8. A Supply Chain Perspective on Price Formation in
   Agri-Food Chains 167
   Gerhard Schiefer and Jivka Deiters

9. Summing Up: New Insights and the Emerging Policy and
   Research Agenda for Addressing Food Price Inflation 187
   Steve McCorriston

Index 193

List of Figures

1.1.	World food price index and UK food consumer price index (2005=100)	3
1.2.	Experience of food and non-food inflation in the UK, Germany, Japan, and the US (1998–2014)	5
1.3.	Experience of food inflation in selected EU Member States (1998–2014)	8
1.4.	UK retail bread and world wheat prices, July 2001–July 2004 (January 2003=100)	12
1.5.	Weekly prices (pence) of a specific bread product across all major UK food retailers (Kingsmill Everyday White Bread)	13
2.1.	Share of household expenditure on food across EU-27	23
2.2.	Price indices for wheat in global and domestic agricultural markets and retail bread prices, 1997–2011 for UK, France, and Poland	25–6
2.3.	Average rates of food inflation across EU Member States, 2000–13	28
2.4.	Average rates of food and non-food inflation across EU Member States, 2000–13	29
2.5.	Coefficient of variation of food and non-food inflation across EU Member States, 2000–13	30
2.6.	Correlation between barriers to competition at retail index and contribution of world wheat prices to retail bread prices	46
2.7.	Correlation between market share of discounters and contribution of world wheat prices to retail bread prices	47
3.1.	Poultry price series for ten EU countries	58
4.1.	Raw milk price in France 1990–2011	69
4.2.	Evolution of prices 2006–09	70
5.1.	Sample retail prices of Radeberger pilsener for different distances between store and brewery (weekly from 2000 to 2001)	105
6.1.	Brand level average weekly prices (euro/litre) for high-quality refrigerated milk in chain A (left panel) and chain B (right panel) (2009–11)	126
6.2.	Brand level average weekly prices (euro/kg) for butter in chain A (left panel) and chain B (right panel) (2009–11)	127

List of Figures

6.3.	Comparison of dairy price indices computed on different data sets (2009–11)	137
6.4.	Comparison of Laspeyres refrigerated milk price indices computed on different data sets (2011)	138
6.5.	Comparison of Laspeyres UHT milk price indices computed on different data sets (2011)	139
6.6.	Comparison of Laspeyres butter price indices computed on different data sets (2011)	140
6.7.	Comparison of Laspeyres yoghurt price indices computed on different data sets (2011)	141
6.8.	Comparison of different refrigerated milk price indices computed on scanner data (2009–11)	142
6.9.	Comparison of different UHT milk price indices computed on scanner data (2009–11)	143
6.10.	Comparison of different butter price indices computed on scanner data (2009–11)	144
6.11.	Comparison of different yoghurt price indices computed on scanner data (2009–11)	145
7.1.	Relationship between producer and consumer prices	157
7.2.	Price transmission (τ)	161
7.3.	Impact of contract costs	162
8.1.	The retail network approach	181

List of Tables

2.1.	Cumulative changes in food prices across EU Member States, 2005–13	24
2.2.	Long-run price transmission elasticities: Retail bread prices w.r.t. world wheat prices	45
3.1.	Estimation of the Vector Error Correction Model: Adjustment of producer and consumer prices to long-run disequilibrium	60
3.2.	Tobit results: Parameter estimates	61
4.1.	Dairy desserts: Descriptive statistics for prices and market shares by categories	73
4.2.	Fluid milk: Descriptive statistics for prices and market shares by categories	74
4.3.	Results of the random coefficients logit model	82
4.4.	Average own-price elasticities between products	85
4.5.	Impact of a 10% decrease in the raw milk price on retail prices	88
4.6.	Regression of pass-through on cost shock variables and product characteristics	90
5.1.	Ranks of regional market shares of beer brands in Germany (2000–01)	106
5.2.	Descriptive statistics of the brands' pricing strategies	113
5.3.	Descriptive statistics of model variables (2000–01)	114
5.4.	Estimation results for spatial and temporal beer brand pricing characteristics (2000–01)	116
7.1.	A typology of contracting costs	153

List of Contributors

Céline Bonnet is a researcher at Institut National de la Recherche Agronomique (INRA) in France within the Toulouse School of Economics (TSE). She specializes in the industrial organization and consumers' behaviour in the agro food chains. Her recent research addresses the assessment of food policies for better health. She has published widely including the *RAND Journal of Economics*, the *Journal of Public Economics*, and the *American Journal of Agricultural Economics*, among others.

Elena Castellari is Researcher at the Dipartimento di Economia Agroalimentare, Università Cattolica del Sacro Cuore, Piacenza, Italy. Her research interests focus on consumer behaviour, industrial organization of the food sector, and on the relationship between food policy and health.

Tifenn Corre is a junior development engineer at Institut National de la Recherche Agronomique (INRA) in France. She specializes in structural econometric modelling applied to the food industry. Her recent work deals with price transmission in food chains and estimation of demand models for different food markets which are developed to assess the impact of food and environmental policies on those markets.

Stephan von Cramon-Taubadel has held the Agricultural Policy Chair at the Georg-August University of Göttingen since 1999. He received BSc and MSc degrees from McGill University and the University of Manitoba, respectively, and a PhD from the University of Kiel in Germany. His research and teaching focus on agricultural policy in the EU and the countries of the Former Soviet Union, and on the measurement of price transmission and agricultural market integration. Stephan has worked as a consultant for the German government, the World Bank, the FAO, and the OECD. From 2000 to 2006 he was editor of *Agricultural Economics*.

Jivka Deiters is currently at the University of Bonn but has been engaged in the EU projects e-trust, Netgrow, and Transparent_Food which were dealing with the utilization of e-commerce developments in international trade relationships, with innovation and learning in food networks, and with transparency in the food chain. She is presently focusing on the Future Internet PPP programme of the EU which promotes the utilization of digital technology for improvements in efficiency, transparency, and sustainability along the food chain and with consumers.

José M. Gil is Professor of Agricultural Economics at the Technical University of Catalonia (UPC) and Director of the Centre for Agro-food and Development Economics (CREDA), which is also responsible for the team of food chain analysis and consumer behaviour. He worked previously as Senior Scientist at the Agricultural Research Service

List of Contributors

(Aragon Government) from 1989 to 2002. His current research focuses on price analysis and the economics of food quality and safety and related policy issues with respect to the consumer, the food industry, and trade.

Thomas Glauben is the Director of the Leibniz Institute of Agricultural Development in Transition Economies (IAMO) and a full professor at the Martin-Luther-University Halle-Wittenberg. He holds a PhD in Agricultural Economics from the University of Kiel. He is a member of the executive board of the Leibniz Association and sits on the editorial boards of several journals. His research focuses on agricultural and food economics.

Islam Hassouneh is a professor of business economics at Palestine Polytechnic University (PPU). After studying economics at Autonomous University of Barcelona (UAB) and Technical University of Catalonia (UPC), he worked as a postdoctoral fellow at the Centre for Agro-food and Development Economics (CREDA). His research interests focus on applied economics, agricultural economics, and price analysis. His work has been published in various journals, including *Energy Economics, Agricultural Economics,* and *Food Policy*.

Carsten Holst is a research assistant in the Department for Agricultural Economics and Rural Development at Georg-August-University Göttingen. He completed his PhD thesis about vertical and horizontal price relations in the European pork sector in 2013.

Tim Lloyd is an associate professor in the School of Economics at the University of Nottingham. His research interests lie primarily in agricultural and food economics, particularly the time series analysis of food and commodity markets and the econometrics of supermarket scanner data. He has undertaken research for a number of national and international organizations including HM Treasury, Department of Food, Environment and Rural Affairs (DEFRA), and the European Commission. He was formerly managing editor of the *Journal of Agricultural Economics* and he is currently a member of DEFRA's Economics Advisory Panel. His research has been published in journals such as the *Journal of Agricultural Economics, European Review of Agricultural Economics, Oxford Economics Papers, Economic Journal,* and the *Oxford Bulletin of Economics and Statistics*. He is currently President Elect and will be President of the Agricultural Economics Society in 2016.

Jens-Peter Loy is a full professor and Director of Department of Agricultural Economics at Christian-Albrechts-University of Kiel. He holds a PhD in Agricultural Economics from the University of Kiel. His research focuses on price transmission and cost pass-through, auctions, and pricing and competition in food retail markets. Recent work has been published in *European Review of Agricultural Economics, Journal of Agricultural Economics,* and *European Journal of Marketing* and *Agribusiness*. He is on the editorial boards of the *European Review of Agricultural Economics* and *Agribusiness*.

Steve McCorriston is a professor in the Department of Economics at the University of Exeter Business School. His research focuses on trade policy and competition issues in the food sector. He was coordinator of the EU FP7 project on the 'Transparency of Food Prices'. His research has been published widely including in the *European Economic Review,* the *American Journal of Agricultural Economics,* and the *Journal of Empirical*

List of Contributors

Finance, among others. He has previously served as associate editor for the *American Journal of Agricultural Economics* and is currently the editor of the *European Review of Agricultural Economics*. He has been a consultant to a number of organizations including the UN FAO, OECD, the UK's Department of Environment, Food and Rural Affairs (DEFRA), and the Department for International Development (DfID).

Wyn Morgan is a professor of economics at the University of Nottingham and has a long-standing research interest in the transmission of prices along food chains. Commodity market functioning and the role of futures markets in stabilizing volatile prices along with an emphasis on the role of competition in shaping the prices consumers pay for the final retail food product have been key areas of enquiry. Funding for his research has come from a number of sources including DEFRA, Foresight, Wincott Foundation, EU FP7 programme, and the Commonwealth Secretariat.

Daniele Moro is Associate Professor in Agricultural Economics at the Dipartimento di Economia Agroalimentare, Università Cattolica del Sacro Cuore, Piacenza, Italy. His research interests focus on food demand and price analysis, agricultural policy, and microeconometrics.

Silvia Platoni is Researcher at the Dipartimento di Scienze Economiche e Sociali, Università Cattolica del Sacro Cuore, Piacenza, Italy. Her research interests focus on political economy, public economics, and microeconometrics.

Vincent Réquillart is a senior researcher at Institut National de la Recherche Agronomique (INRA) in France. He is deputy head of the Groupe de Recherche en Economie Mathématique et Quantitative (GREMAQ) unit, within the Toulouse School of Economics (TSE). His research focuses on agricultural economics and industrial organization applied to the food industry. His recent research relates to the economics of the food chain, in particular assessing food policies for better health. He serves on the editorial board of the *European Review of Agricultural Economics*.

Gerhard Schiefer has held professorships in business management at the universities of Kiel, Hohenheim, and (currently) Bonn, and served as founding director of the International Center for Food Chain and Network Research at the University of Bonn. He has been co-initiator of the European Technology Platform (ETP) Food for Life where he is member of the board and chairs the working group on Food Chain Management. He is editor of the *International Journal on Food System Dynamics* and was founding president of the European Federation for IT in Agriculture, Food and the Environment (EFITA). He chairs the International Center for Management, Communication and Research (CentMa).

Paolo Sckokai is Associate Professor in Agricultural Economics at the Dipartimento di Economia Agroalimentare, Università Cattolica del Sacro Cuore, Piacenza, Italy. His research interests focus on food demand and price analysis, agricultural policy, and the industrial organization of the food sector. In the past he has worked for the OECD Agricultural Directorate. From 2007 to 2012 he has served as editor of the *European Review of Agricultural Economics*.

Teresa Serra joined the Agricultural and Consumer Economics Department at the University of Illinois as an associate professor in 2014. She worked as a researcher at

List of Contributors

the Center for Agro-food Economy and Development (CREDA-UPC-IRTA), Barcelona, Spain from 2004 to 2014. She has authored more than fifty papers that have been published in international peer-reviewed journals. She has also authored several book chapters and presented over fifty papers in professional meetings. Her main research interests include price analysis, time-series econometrics, microeconomic analysis, production, and efficiency economics.

Johan Swinnen is Professor of Economics and Director of the LICOS Centre for Institutions and Economic Performance at the University of Leuven (KU Leuven). He is Senior Research Fellow at the Centre for European Policy Studies (CEPS) and Visiting Scholar at the Centre for Food Security and the Environment, Stanford University. He has published widely on institutions and political economy of agricultural development and food policy. His books include *Quality Standards, Value Chains and International Development, Political Power and Economic Policy, Global Supply Chains, Standards and the Poor, Foreign Direct Investment and Human Development, Private Standards and Global Governance,* and *From Marx and Mao to the Market.*

Anneleen Vandeplas is an economist at the European Commission and senior research affiliate with LICOS Centre for Institutions and Economic Performance at the University of Leuven (KU Leuven). Previously she was a visiting professor at the University of Hasselt and visiting researcher at the International Food Policy Research Institute (New Delhi). Her work has been published in international journals including the *World Bank Economic Review, Journal of Development Studies,* and the *Journal of Agricultural Economics and World Development.*

1

Introduction

Steve McCorriston

The world commodity price 'spikes' of 2007–08 and 2011 generated considerable interest in the characteristics of price dynamics and the implications and potential solutions to commodity price volatility. Against the background of relatively low and stable prices for agricultural commodities traded on world markets over the 1990s and first half of the 2000s, considerable effort has been devoted to understanding the causes of the price shocks that were experienced between 2007 and 2011, the key contributory factors relating to poor harvests in key regions, low stocks, biofuels, high demand in emerging economies, financialization of commodity futures markets, and the insulation effect of trade policies. Various insights into these factors can be found in Wright (2011), Baffes and Haniotis (2010), Irwin and Sanders (2011), Martin and Anderson (2011), and Townsend et al. (2011), among others. Although at the time of writing (February 2015) prices on world agricultural markets had fallen by 25 per cent from their peak in 2011, these declines underpinned the volatile nature of commodity prices: the expectation is that world prices will likely rise in the future due to the strength of demand associated with population and income growth outweighing the supply-side effects of productivity growth, while climate change impacts will potentially further hinder supply potential in the absence of mitigation (see, e.g., Nelson et al., 2010). In sum, despite the recent falls from the peaks of 2007–08 and 2011, concerns over the functioning of world commodity markets will likely persist, with a consensus that the era of low agricultural prices on world markets has likely passed and that price volatility with occasional spikes will be a feature of agricultural markets over the next couple of decades. In this context, a better understanding of what drives commodity prices continues to attract attention, and how to address commodity price volatility remains an important issue for stakeholders and policymakers.

However, raw agricultural commodities traded on world markets (to which most of the analysis of the commodity price spikes reported above relates) are not the same thing as domestic food prices that consumers typically face. This is true of both developed and developing countries.[1] There are many reasons why the behaviour of domestic commodity prices may differ from world prices: exchange rates, the use of trade policy instruments (in the case of importing countries, this would involve varying applied tariffs and, in the exporting country cases, imposing export taxes), the use of safety nets, and the use of public subsidies to keep domestic food prices low.

Another reason why domestic food prices behave differently from world agricultural prices is that in the transformation of agricultural commodities into (processed) food products, raw commodities are combined with other inputs before being distributed and sold at the retail stage. In this case, the price of the raw agricultural commodity is not the same as the price of the processed food product sold in supermarkets and other retail outlets. To highlight this difference, in Figure 1.1 we contrast the experience of the retail food price index in the UK with the world commodity price index for the period 1997–2014. The most apparent observation from this figure is how different the price dynamics are between these two distinct stages in the food chain, with the world price index for raw commodities being characterized by volatility and the (UK) retail consumer price index for food being relatively stable. In broad terms, there are two reasons that may give rise to the difference in the nature of price behaviour between retail food markets and world agricultural markets. First, raw agricultural commodities tend to account for a relatively small share of the input cost of the final retail food product, and this share has been declining over time; even for products sold at retail that undergo a relatively small degree of processing, the costs of packaging and distribution can account for a significant share of the costs of the final product. Second, the activities that comprise the food chain (most obviously, food processing and food retailing) may influence the characteristics of retail price dynamics and, at least as far as domestic producers are concerned, also the behaviour of prices for upstream (farm) commodities.

Taking these two perspectives together implies that despite the wide-ranging focus in recent years on the causes of world commodity price spikes, the behaviour of domestic prices may differ from that of world prices for the same commodities. Underlying the relationship between world and domestic prices is the process of price transmission. There are two dimensions to the price transmission process: how the behaviour of world prices translates into

---

[1] As IMF (2011) reported, the change in domestic prices around the time of the 2007–08 crisis on world commodity markets was not only different in magnitude but in some cases also in direction.

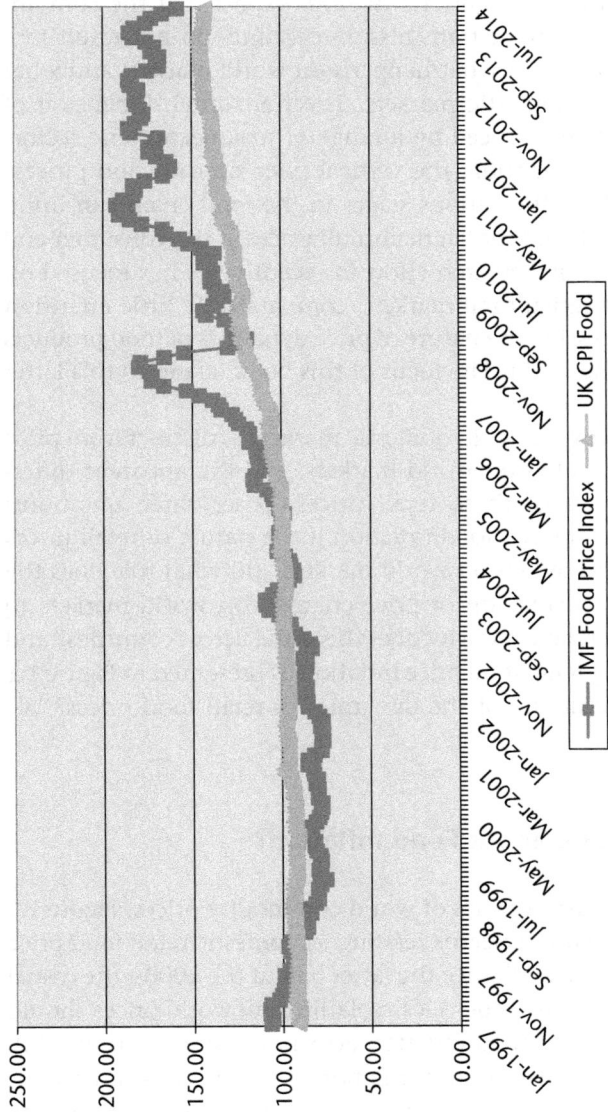

Figure 1.1. World food price index and UK food consumer price index (2005=100)
Source: OECD

price behaviour for identical commodities in the domestic market relates to *horizontal* price transmission; and the relationships between upstream raw commodity prices (whether world prices or domestic farm prices) and the processed food product sold at retail relates to *vertical* price transmission. From the evidence presented in Figure 1.1, we can expect that the combination of these two aspects of price transmission certainly do not result in a one-for-one relationship between what happens on world markets and what the resultant impact is on domestic markets. The horizontal dimension of price transmission may be influenced by a range of macroeconomic factors and government policy intervention; the vertical price transmission process involves understanding how the various stages in the food chain can influence the behaviour of retail food (not agricultural) prices at the consumer end of the food chain. While most research effort in recent years has focused on price dynamics on world agricultural markets, comparatively little attention has been allocated to addressing the nature of price dynamics of food products sold at the retail end of the chain. The focus of this book relates to this latter dimension.

The focus on food price dynamics in domestic markets as distinct from price behaviour for raw commodities on world markets, and the apparent differences in the characteristics of these retail prices, poses three questions: (i) should we care about domestic food inflation if the nature of retail prices is so distinct from price behaviour on world markets? (ii) what role does the food sector play in the transmission of price changes on world markets to domestic food inflation at retail, and why does this differ across countries? and (iii) do the data underlying retail food price inflation as presented in Figure 1.1 accurately reflect the true nature of the dynamics of retail food prices? We address these issues in turn.

1.1 Why Should We Care about Food Inflation?

Given the more volatile characteristics of world commodity prices, Figure 1.1 does not give full justice to the concerns relating to domestic retail food price inflation that have been expressed since the latter half of the 2000s; the casual observation that retail prices are (much) less volatile than world prices should not lead to the conclusion that food inflation does not matter. In Figure 1.2, we present data for food and non-food inflation for two European Union members (the UK and Germany), Japan, and the US. It is clear from the data that steep rises in domestic retail food price inflation coincided with the commodity price spikes that were experienced on world markets, and that food price inflation was certainly more volatile (and, on average, higher) than non-food inflation. Further, although this volume focuses on the retail price

Introduction

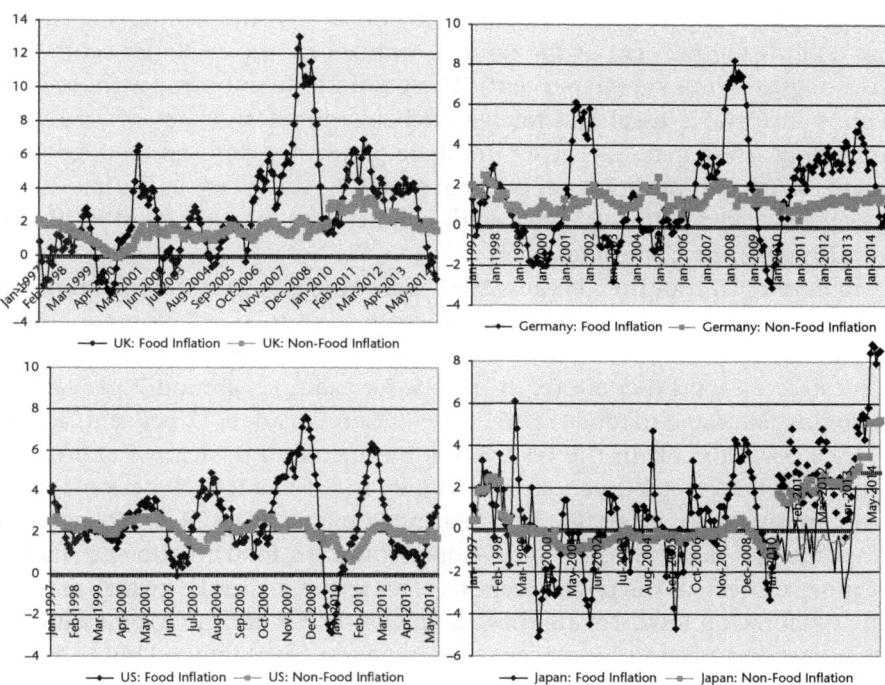

Figure 1.2. Experience of food and non-food inflation in the UK, Germany, Japan, and the US (1998–2014)
Source: OECD

dynamics in the EU, clearly the experience of food inflation compared with non-food inflation in the EU is not untypical of developed countries, as the data for Japan and the US indicate.

The most obvious concern with food inflation is that high food prices impact on the poorest income groups. In the context of recent events on commodity markets, the World Bank reported that due to the commodity crises of 2007–08, the number of new poor increased by 108 million (Ivanic and Martin, 2009); Ivanic et al. (2011) estimated a rise in new poor of 44 million associated with the 2011 price spike. The impact in developing countries is particularly strong where a large proportion of income is spent on food. Aside from the direct human cost, high food prices have wider implications both in terms of how governments deal with reducing the impact of high food prices and the potential implications for political stability. As noted above, domestic prices can behave differently from world market prices. Evidence from the World Bank (2009) illustrates a wide range of responses by governments to events on world markets: in a sample of fifty-eight countries, only 15 per cent of countries had no response, the other countries employing export restrictions (20 per cent of the sample), controlled prices (30 per cent), and reduced taxes on food (40 per cent). These actions do not come without cost, with considerable fiscal and foreign exchange implications associated with providing safety nets. Yet, given the share of income spent on food, governments also have concerns over political stability—the events in the late 2000s being associated with food price riots, conflict, and, in the case of Haiti, the overthrow of the government.

In developed countries, because the share of income spent on food is much smaller, in aggregate the potential implications of food inflation are potentially less severe. As is well known, as incomes rise, a smaller share of income is allocated to food expenditure. In the US, for example, around 7 per cent of income is allocated to food expenditure; in Germany, it is 12 per cent; and in France, 14 per cent. In the relatively lower-income EU Member States, the share of food in total expenditure is higher: in Estonia it is 20 per cent, with the corresponding shares for Lithuania and Romania being 24 and 28 per cent respectively.[2] However, even in developed countries, food inflation can have a regressive effect, as the poorer sections of society still spend a greater share of their income on food. Taking the example of the US, for the middle quintile income group, around 13 per cent of income was spent on food but this rises to around 36 per cent for the lowest quintile.

Food inflation also has a macroeconomic dimension, and with the high levels of domestic food inflation that have been experienced in recent years, the issue arises as to what, if anything, monetary authorities should do about it? Despite the obvious impact that rising food prices will clearly have on consumer and household budgets, it is less obvious what the response of macroeconomic policymakers should be. This is due to the fact that the concerns of macroeconomic policy focus on 'core' inflation, which generally is the headline inflation rate with food (and energy) prices taken out. It is this 'core' rate which is seen to be important for setting macroeconomic policy. Why, then, is it the case that with rising commodity and food prices, reflected in food and headline inflation across many countries, macroeconomists set issues of food price inflation aside when it comes to anti-inflationary policy? The answer lies in the observation that commodity and food price shocks and the resultant volatility in domestic retail food inflation are seen to be transitory in nature and may not affect 'inflationary expectations'. The key link here is the so-called propagation mechanisms or second-round effects, i.e., the way in which food price inflation will affect non-food (or 'core') inflation.

---

[2] The data reported here is from the USDA and relates to food expenditure at home, excluding expenditure on alcohol and beverages.

Specifically, if food price inflation is high and persistent, then this will affect wage inflation across the whole of the economy (say through bargaining by labour unions). The emphasis here is not just on the level of prices, but also the persistence of the change: if the change in prices is long-lasting, then food price inflation will matter for macroeconomic policy. But if food price inflation is transitory and non-persistent, then the propagation mechanism will be weak and food price inflation should be set aside by macroeconomic policy-makers. Further, targeting inflation in response to transitory events will only result in output volatility. In sum, more stable headline inflation targets can be met by the monetary authorities focusing on the underlying ('core') level of inflation.

In light of recent events on global commodity markets and domestic food (and energy) price inflation, there has been some research addressing these issues. Cecchetti (2007) notes that excluding food from core price inflation is only justified if the long-run mean of food and non-food inflation are equal; if they are not, then this may justify greater focus on food inflation by monetary authorities. In a recent empirical study of this issue, Cecchetti and Moessner (2008) conclude that core inflation does not converge to headline inflation (highlighting the transitory nature of food price inflation) and that higher global commodity prices (as reflected in domestic food price inflation) have not generated strong second-round effects on core inflation. IMF (2011) address these issues across a large sample of advanced and emerging/developing economies, while Walsh (2011) focuses on various measures of persistence and transmission between food and non-food price inflation, again across a large sample of advanced and emerging/developing countries. These studies broadly lend support to the exclusion of food price inflation from 'core' measures, confirming that food price inflation does not result in significant second-round effects. Setting aside food price inflation may not be the optimal policy in all circumstances, however: Anand and Prasad (2010) show that in economies where food expenditure shares are high and in the presence of financial frictions (e.g., borrowing constraints), the optimal policy may be for the government to focus on headline inflation (i.e., inclusive of food price inflation) in stabilizing prices and output in the economy.

In sum, food price inflation has important consequences, though these differ between developed and developing countries. High food prices impact on the poorest sections of society and (particularly where the second-round effects are strong) have potential macroeconomic implications, the impact of these effects being particularly severe for developing countries. However, even in developed countries, important questions arise when examining the differences in the food inflationary experience and the causes and potential impacts of these differences; we turn to these issues next.

1.2 Why Does the Experience of Food Inflation Differ?

In broad terms, it is straightforward to rationalize why the experience of domestic food inflation differs across countries. As discussed, in developing countries, given the overall importance of food in expenditure, the fact that the 'persistence' effect of food prices may impact on wage setting and the potential political stability implications, governments may prioritize actions to moderate the impact of price spikes on world markets. In developed countries, food inflation may also differ due to openness to world markets, exchange rate movements, differences in trade policy regimes, and so on.[3] However, even within the EU, where countries have common trade policies and—at least for a subset—common exchange rates, the experience of food inflation has differed. This is evident from Figure 1.3, which shows food inflation for several of the more established Member States of the EU.

Of course, while differences associated with degrees of openness to world markets and exchange rate regimes (e.g., the UK is not a member of the

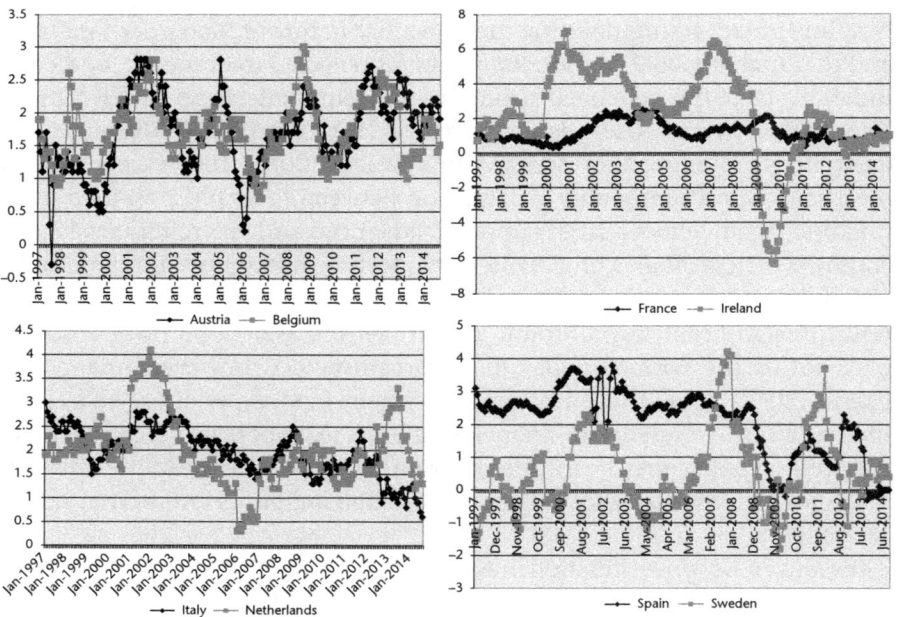

Figure 1.3. Experience of food inflation in selected EU Member States (1998–2014)
Source: OECD

---

[3] These issues relate to the literature linking globalization to cross-country experience of inflation. See, e.g., Pain et al. (2008).

Introduction

Eurozone, while other countries reported in Figure 1.3 are) may influence the domestic food inflation experience, attention has turned to differences in the structure of the food sector across EU Member States and, by extension, how the food chain functions. These issues were highlighted in an earlier report from the EU Commission (Bukeviciute et al., 2009) which noted differences in the EU food sector relating to concentration in the food manufacturing and retailing stages, the role of discounters, the proliferation of private labels, and so on with related concerns highlighted by the EU Commission with specific reference to the functioning of the food supply sector across the EU (EU Commission, 2009). The European Central Bank also highlighted the potential significance of the retail sector (i.e., the high levels of concentration) and discounters (i.e., the potential pro-competitive effect associated with their high market shares in some EU Member States) in potentially impacting on the variation in the food inflation experience across the EU (European Central Bank, 2011). In broad terms, the links between the events on world markets and domestic food inflation, and the structure and functioning of the food sector in different EU Members States, relates to the price transmission process, specifically the concerns relating to linking competition throughout the food supply chain with price dynamics at the retail level. Consistent with the overall concerns about competition and the behaviour of food prices, a report by the European Competition Network (2012) provided an audit of anti-trust investigations and sector reviews that had recently been directed at the food sector in EU Member States.

Given—as we have noted above—that world prices for raw commodities are not the same as retail prices for food products, it is clear that the characteristics of the food sector could impact on the price transmission process. Specifically, except in the case of perfect competition, complete markets and perfect information, which would render the economic impact of the food sector on price transmission innocuous, the structure and functioning of the food sector would clearly have some impact on the transmission of world price shocks on domestic retail food inflation. Yet the functioning of the food sector is complex and the links between price transmission and competition not always obvious. Competition in the food sector relates to a number of different dimensions covering seller and buyer power, concentration at several stages in the food chain, the nature of the links between successive stages, the proliferation of private labels, the impact of the rising market share of discounters, the emergence of buyer groups, and so on, all of which indicate that just looking at firm numbers (i.e., concentration ratios) does not necessarily reflect anything precise about anti-competitive behaviour in the food chain and, by extension, the possible links between the functioning of the food chain and price transmission. But clearly, differences in the functioning of the food sector do exist across the EU Member States and, in principle, may have some impact on the price transmission process and, in turn, contribute to explaining the differences in food price inflation across EU Member States. In sum, understanding how the food chain functions and how this functioning of the food chain impacts on the characteristics of retail food price dynamics is a priority for addressing food inflation. This is the main issue that is addressed in the contributions to this volume.

Finally, it is worth noting why price transmission matters. At one level, it underpins the food inflation process as price changes are fed through the food system from world or upstream markets through to retail. Even if—due to the inherent volatility of retail food inflation—this does not lead to monetary authorities playing an active role in targeting food inflation as discussed above, the price transmission process still matters. First, food inflation matters because it still has a strong impact on the cost of living and is also particularly regressive, affecting the poorest most. Second, with retail prices being apparently more stable than upstream raw commodity prices, the burden of price adjustment will take place at other stages of the food sector: that is, prices at the domestic farm gate will likely be more volatile than prices in retail markets, which poses the adjustment elsewhere in the food sector and which has implications for the use of (or the demand for) agricultural and trade policies. Finally, if the features of the food sector mean price transmission is imperfect, then understanding the mechanism that causes this is important. If it is the case that food retailers and manufacturers are adjusting their markups, this implies an additional aspect of distributional effects throughout the food chain. In its starkest form, if retail food prices do not fall as much as prices on world markets, does this mean that markups in the food sector are increasing, and is it the absence of competition that causes this? These issues, contingent on the factors which drive the price adjustment process, give rise to concerns about 'fairness' throughout the food sector.

1.3 Data on Food Prices

The data underlying Figures 1.1 and 1.2 relate to data sources most readily available to researchers and which are most readily reported. Data on prices on world markets are available for specific commodities and in index form for groups of commodities (or as an aggregate), while the data on retail food prices relate to the consumer price index published by the national statistical offices (in this case, the UK Office for National Statistics) which is therefore accessible for a wider range of countries. This consumer price index for food reports monthly data for a basket of food products at retail; other more disaggregated data for specific categories (e.g., bread and bread products, chocolate, and confectionery) are also available. The most apparent feature of the data

Introduction

presented in Figure 1.1 suggests that retail food price inflation, though it has risen in recent years, is more stable than prices for commodities traded on world markets. Data for sub-categories (bread versus wheat) would show the same features. While there are some obvious reasons that may underpin this relative stability (as noted above, raw commodities are a small share of the processed food product sold at retail, and the characteristics of the food chain may impact on the price transmission process), the question also arises as to whether retail food price indices that are readily accessible to researchers and most regularly reported are representative of price dynamics at the retail level.

Research activity using more disaggregate price data to explore in more detail the dynamics of price inflation has been a growth area in recent years. There have been two dimensions of this. First, the Inflation Persistent Project explored features of retail prices across several Eurozone countries using data from national statistical sources. These are the source data that underpin the reported monthly consumer price indices. This project—the details of which are summarized in Dhyne et al. (2006)—highlighted several dimensions of retail price dynamics that would not be obvious from the monthly and aggregated consumer price indices, including how often and how much individual prices do change. This strand of research highlights considerable heterogeneity across sectors and across countries in the price adjustment processes. While insightful, it is still limited to monthly data collected from surveys and still may not give a full picture of retail (food) price dynamics.

The second source of data which casts light on price dynamics relates to scanner data. These data are high frequency (typically weekly) and are available at the product-specific (or unique product code—UPC) level. These scanner data give a very different perspective on retail prices that would not be apparent from the data reported by the national statistical sources and underlying food price inflation indices. Eyeballing scanner data suggests that food prices are not as stable as commonly construed from the monthly aggregates and that the prices of even identical products sold across different retailer chains indicate that the nature of price dynamics is more complex than investigation of the monthly inflation aggregates would suggest.

This is apparent from Figure 1.4. Here we report the monthly retail bread price index for the UK, which is sourced from publically available sources (we restrict the time period covered here to July 2004–July 2007 to be consistent with the data reported in Figure 1.5). To underpin the relative 'stability' of retail bread prices, we also report in Figure 1.4 world wheat prices, the differences in price dynamics between these price indices at the two ends of the supply chain being consistent with the discussion above that retail food prices are apparently more stable than world raw commodity prices. However, in Figure 1.5, we report scanner price data for the same UPC-specific bread sold in the five major retailers in the UK. There are two things to note from the

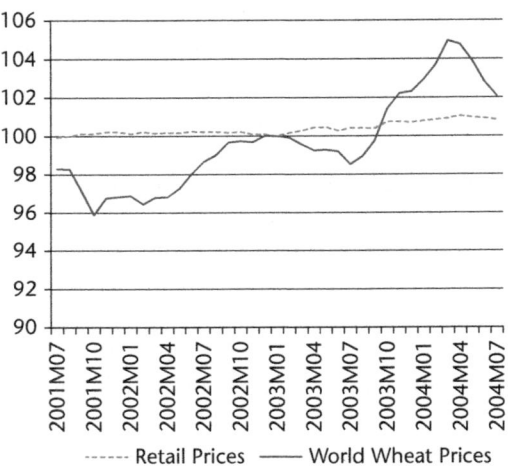

Figure 1.4. UK retail bread and world wheat prices, July 2001–July 2004 (January 2003=100)
Source: UK Office for National Statistics

comparison. First, scanner prices for UPC-specific bread products seem to bear little resemblance to the monthly aggregate price indices that are publically available and widely referred to. Second, price dynamics for the same UPC product vary across different retailers. This observation of the behaviour of scanner prices for food products carries over to all other food categories both in the UK and other countries.

While, taken at face value, retail food price dynamics using scanner data appear to be substantially different from the monthly aggregates commonly reported, there are a number of challenges in characterizing scanner prices. First, since many of the price changes may be products on sale at any point, should we distinguish between temporary reductions in price (i.e., sales prices) and, in turn, how does the researcher define a 'sale' if this is not flagged in the initial data? Second, what is the feature of prices that characterizes firms' strategies? Is it confined to non-sale prices or to some other dimension, such as the notion of reference prices introduced by Eichenbaum et al. (2011)? These issues are important, and while much of the evidence to date relates to the US (see Klenow and Malin (2011) for a review of the range of insights that can arise from micro data sources), there are a number of additional insights that relate to scanner data from the food sector across the EU. These include the heterogeneity in retail price behaviour across retail chains (Lloyd et al., 2014), the fact that retail price behaviour can vary even within a national chain depending on the outlet (see Castellari et al., this volume), and, by extension, that the process of price transmission may vary depending

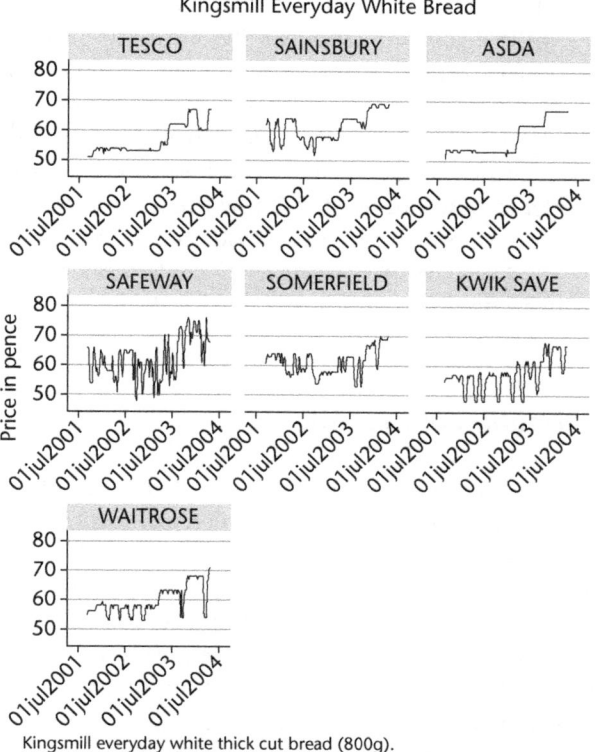

Kingsmill everyday white thick cut bread (800g).
Graphs by retail chain. Sample period 08 September 2001 to 17 April 2004.

Figure 1.5. Weekly prices (pence) of a specific bread product across all major UK food retailers (Kingsmill Everyday White Bread)
Source: Lloyd et al. (2014)

on whether the data refer to national brands or private label products (see Loy et al., 2015). Scanner data have also been used to evaluate the incidence of taxes on food products that also allow for the potential linkages between food retailers and food manufacturers (see Bonnet and Réquillart, 2013).

The availability of scanner data offers considerable potential for providing a clearer perspective on the dynamics of retail food prices and, by extension, the functioning of the food supply chain. However, in addition to the research challenges in making sense of scanner data, there are a number of other issues to be aware of. First, scanner data sources may only provide data on prices, not necessarily on quantities as well. This determines the range of questions that can be addressed with scanner data; for example, in the absence of quantity data, estimating structural models and deriving product- (UPC-) specific elasticities is not always possible, the study of the French sugar sector by Bonnet and Réquillart (*ibid.*) being an exception in this case. Secondly, scanner data covering different EU countries are not homogeneous. Some only report national data, while others distinguish by location; others differentiate between retail outlets, while still others provide a single 'national' price. Finally, there is the issue of the financial cost of acquiring these data, which are often from private sources. Despite these issues, the insights from scanner data sources have considerable potential to address food price dynamics at the retail stage in order to better understand how the food sector functions, how price transmission develops through the food chain, and the heterogeneous nature of price responses. In short, scanner price data can increase the transparency of the food chain and of how and why it differs across the EU.

1.4 The Contribution of this Volume

The chapters in this volume address the issue of retail food price dynamics across the EU by examining the links between price transmission and the functioning of the food supply chain in different EU Member States. They draw on a variety of data sources, employing publically available (monthly) price indices and scanner data covering retail prices in selected EU Member States. The research reported in this volume relates to an EU-funded project on the 'Transparency of Food Prices' which involved a collaboration of researchers across ten EU Member States.[4] The research reported here involves a mix of theory and empirical analyses and, while principally focusing on the economic analyses of retail prices and the characteristics of the food sector in the EU, also reports insights from a supply chain perspective more closely associated with business and management.

In Chapter 2, Lloyd et al. address two broad issues. First, they overview the experience of food inflation across EU Member States, highlighting the different experiences associated with food inflation (both in terms of average levels and variability) and also how the experience of food inflation has differed from non-food inflation. Second, since the issue of price transmission lies behind events on world commodity markets and the behaviour of retail food prices, they review the theoretical and empirical literature on price transmission. A key feature running through research in this area and the policy debate is the concern that market structure in the food sector is a principal determinant of how retail prices respond to upstream shocks, and the emphasis in the literature review is to highlight the dimensions of

---

[4] 'Transparency of Food Prices' (TRANSFOP) project funded by the European Commission, Directorate General Research-unit E Biotechnologies, Agriculture, Food. Grant Agreement No. KBBE-265601-TRANSFOP. See also <http://www.transfop.eu>.

Introduction

competition in the food sector which may determine the extent of price transmission between stages in the food chain. Finally, drawing on related work on factors that determine price transmission across the EU, they show that the varied experience of food inflation is indeed likely to be correlated with the extent of competition in the food sector in EU Member States.

In Chapter 3, Hassouneh et al. address the comparative experience of price transmission across the EU, specifically by assessing the nature and degree of price transmission for nine different commodity chains across ten EU Member States. They show that retail prices respond less than upstream producer prices but reject the overall presence of asymmetric price adjustment (i.e., that the price transmission effect is contingent on whether upstream prices are rising or falling), a concern that has often been linked to the exercise of market power in the food sector and which has been of concern to policymakers. Finally, they bring together the price transmission estimates and look for common factors that may explain why price transmission varies across both sectors and countries. Their results suggest that production share within the EU, the extent of specialization (both in terms of export and production specialization), and the relevance of vertical contracts all impact on the price transmission process.

These two chapters largely rely on monthly data relating to some 'food' or 'commodity' aggregate to address retail food price dynamics; as discussed above, the use of high-frequency scanner data has the potential to supply more detailed insights. Chapters 4 to 6 present research that highlights the range of insights that can be obtained from scanner data.

Bonnet et al. (Chapter 4) use scanner data relating to the French dairy industry where, at retail, they have data covering 20,000 households over a four-year period. Based on this, their analysis contains several attractive features: since they have both price and quantity data, they can estimate a structural econometric model that allows them to derive product-specific demand elasticities; since they have estimates of costs, they use this information to construe the form of vertical relations that characterize pricing behaviour between food manufacturers and retailers; finally, based on these two components, they estimate the price transmission effect in the French dairy desserts and fluid milk sectors following a decrease in the raw milk price. Their results show considerable heterogeneity in price transmission in these two sectors; for some products there is over-shifting (price transmission greater than unity) and on others under-shifting (price transmission less than unity). Perhaps more insightful is their identification of the mechanisms that may lie behind these price transmission effects, including the nature of vertical contracts, the distinction between national brands and private label products (which can be interpreted as relating to the food retailers' control of the supply chain), and the adjustment in the food industry markup as important determinants of the price transmission effect (which confirms the importance of this mechanism as highlighted in the literature review in Chapter 2).

Not all scanner sources provide the same information as it depends on the source of the data, the consequence of this being that there is a potential barrier (at present) to carrying out identical comparative work for different EU Member States. However, the compensating effect of this is that alternative data sources can provide insights into different dimensions of the functioning of the EU food sector and how it relates to price dynamics. In Chapter 5, Loy and Glauben employ scanner data for the German beer market to analyse the spatial and temporal pricing associated with brand loyalty and seasonal pricing. These issues relate to consumer search and the pricing strategies of retailers, addressing the notion that with well-known brands and in periods of peak demand, price promotions may be more intense. Their results highlight the nature of spatial and temporal pricing strategies in the German beer market, these insights from brand-specific scanner data not being possible with the use of average national prices.

The detailed information associated with high-frequency scanner data also has implications for the measurement of inflation. This issue is addressed by Castellari et al. in Chapter 6 using scanner data from the Italian dairy sector. Casual inspection of these data shows that price dynamics at the retail stage are contingent on whether the product is a national brand or private label, where the product is sold (hypermarket, supermarket, or convenience store) and, even for identical products, across retail chains. This therefore casts some doubt as to what the average (national) price that enters into the consumer price index reflects. To address this issue, they derive retail price indices that can accommodate the features of scanner data and compare these price indices with the nationally reported price index. They show there are substantial differences in the measurement of retail price changes depending on whether the data refer to nationally available monthly aggregates or high-frequency detailed scanner data. This is important, not just because it reflects additional insights relating to the dimensions of retail price behaviour across commodities, retail chains, and retail outlets, but also because several national statistical authorities are giving consideration to the use of scanner data to more accurately measure inflation.

In Chapter 7, Swinnen and Vandeplas provide new theoretical insights into the price transmission process. As outlined in the review of the theory on price transmission in Chapter 2, the links between the features of the food sector and the price transmission effect highlights aspects of competition and market structure (e.g., how the industry markup changes at each stage, the role of vertical restraints, and so on). Swinnen and Vandeplas focus on different issues that relate to the functioning of food supply chains, specifically the role of contracting, where contracts can involve levels of investment by the buyer and where contracts may or may not be perfectly enforced. They show that the use of contracts is not only important in determining rent distribution between stages in the food chain, but also influences the price transmission process. In this set-up, the patterns of price transmission may be complex and can be discontinuous in nature (e.g., for a range of cost changes, retail prices may not change at all, but in a different range, there may be step change in retail prices following a change in costs). This contribution highlights that retail price dynamics will depend on a wide range of factors that characterize relations in the supply chain, including how aspects of vertical coordination are monitored, and that price transmission is not solely contingent on the nature of competition in the food sector which most theoretical and empirical work has relied upon to date.

The contributions to this book as summarized above draw on new insights from economists on retail price dynamics; but economists do not have a monopoly on insights into how the food sector functions. In Chapter 8, Schiefer and Deiters outline a supply management approach to the functioning of the food chain and highlight the key ingredients in the methodology that is applied. So, while economists rely on theory to derive appropriate hypotheses and on access to data to specify and estimate econometric models, the supply management approach relies on focus groups, networks, and case studies to inform scenario analysis of factors that may drive changes in the food sector. The insights provided in this chapter outline a different perspective to promoting transparency in how food chains actually function.

Taken together, the contributions to this volume aim to provide new insights into retail price dynamics and, more generally, how food prices at the retail stage are linked to events in upstream and world markets. While the focus has been on the EU food sector and food price dynamics across EU Member States, we hope that the theoretical and empirical work (including insights into the use of high-frequency scanner data) will prove useful in furthering research in this area, not just in the EU but further afield. While the commodity price spikes have diminished for now, future concerns over the likely trends in world commodity markets coupled with the ongoing changes in the food sector (particularly industry consolidation and the growth of retailer chains) suggests that continued research on the functioning of the food chain, and the extent and nature of competition in the food sector (both in Europe and elsewhere), will continue to be an important area of enquiry. It is an issue that will be of increasing importance to stakeholders and policymakers as they endeavour to make sense of how events on world markets are linked to domestic welfare issues, and how competition throughout the food sector as well as the different dimensions of competition matter in influencing price transmission. In sum, the issues covered here hold considerable promise for further research and indicate where such research can have a direct impact on policy options (covering competition and regulation, through to agricultural policy) that will be directed at addressing concerns in the food sector.

In closing, I would like to acknowledge the contributors to the TRANSFOP project, especially the contributors to this volume. On behalf of my TRANSFOP colleagues, we are grateful to the EU Commission for financial support and our two project officers, Agata Piendiaz and Dirk Pottier, for their support throughout the project. Finally, a key feature of the TRANSFOP project was to encourage close cooperation between stakeholders (i.e., the policy and industry representatives) and researchers and we are grateful for the time they have given over the project, helping us frame the issues and disseminate the results of the research on the food pricing issues over the duration of the project.

References

Anand, R. and E. S. Prasad (2010) 'Optimal price indices for targeting inflation under incomplete markets'. Cambridge, MA: NBER Working Paper No. 16290.

Baffes, J. and T. Haniotis (2010) 'Placing the 2006/08 commodity price boom into perspective', Policy Research Working Paper No. 5371. Washington DC: World Bank.

Bonnet, C. and V. Réquillart (2013) 'Impact of cost shocks on consumer prices in vertically-related markets: the case of the French soft drink industry', *American Journal of Agricultural Economics*, 95: 1088–108.

Bukeviciute, L., A. Dierx, and F. Ilzkovitz (2009) 'The functioning of the food supply chain and its effect on food prices in the European Union'. Brussels: European Economy Occasional Papers 47.

Cecchetti, S. (2007) 'Core inflation is an unreliable guide', VoxEU, 1 March 2007.

Cecchetti, S. and R. Moessner (2008) 'Commodity prices and inflation dynamics', *BIS Quarterly Review*, 55–66.

Dhyne, E., L. J. Àlvarez, H. L. Bihan, G. Veronese, D. Dias, J. Hoffmann, N. Jonker, P. Lunnemann, F. Rumler, and J. Vilmunen (2006) 'Price setting in the Euro Area and the United States: evidence from individual consumer price data', *Journal of Economic Perspectives*, 20: 171–92.

Eichenbaum, M., N. Jaimovich, and S. Rebelo (2011) 'Reference prices, costs and nominal rigidities', *American Economic Review*, 101: 234–62.

EU Commission (2009) '*A Better Functioning Food Supply Chain in Europe*', Communication from the Commission to the European Parliament, the Council and the European Economic and Social Committee and the Committee of the Regions. Brussels: COM (")() 591 Final.

European Central Bank (2011) 'Structural features of the distributive trades and their impact on prices in the Euro Area', European Central Bank, Occasional Paper Series, No. 128. Frankfurt: European Central Bank.

European Competition Network (2012) *ECN Activities in the food sector: report on competition enforcement and market monitoring activities by European competition authorities in the food sector*. Brussels: European Commission.

IMF (2011) *Global Economic Prospects Report, 2011*. Washington DC: International Monetary Fund.

Irwin, S. H. and D. R. Sanders (2011) 'Index fund, financialisation and commodity futures markets', *Applied Economic Perspectives and Policy*, 33: 1–31.

Ivanic, M. and W. Martin, (2009) 'Implications of higher global food prices for poverty in low-income countries', *Agricultural Economics*, 39: 405–16.

Ivanic, M., W. Martin, and H. Zamman (2011) 'Estimating the short-run poverty impacts of the 2010–11 surge on food prices', Policy Research Working Paper 5633. Washington DC: World Bank.

Klenow, P. J. and B. A. Malin (2011) 'Microeconomic evidence on price-setting', in *Handbook of Monetary Economics*, ed. B. Friedman and M. Woodford, pp. 231–84, Amsterdam: North-Holland, Elsevier.

Lloyd, T. A., S. McCorriston, C. W. Morgan, E. Poen, and E. Zvou (2014) 'Retail price dynamics and retailer heterogeneity: UK evidence', *Economic Letters*, 124: 434–8.

Loy, J.-P., T. Holm, C. Steinhagen, and T. Glauben (2015) 'Cost pass through in differentiated product markets: a disaggregated study for milk and butter' *European Review of Agricultural Economics*, 42: 441–71.

Martin, W. and K. Anderson (2011) 'Export restrictions and price insulation during commodity price booms', Policy Research Working Paper No. 5645. Washington DC: World Bank.

Nelson, G. C., M. W. Rosengrant, A. Palazzo, I. Gray, C. Ingersoll, R. Robertson, S. Tokgoz, T. Zhu, T. B. Sulser, C. Ringler, S. Msangi, and L. You (2010) *Food Security, Farming, and Climate Change to 2050*. IFPRI Research Monograph. Washington DC: International Food Policy Research Institute.

Pain, N., I. Koske, and M. Sollie (2008) 'Globalisation and OECD consumer price inflation', *OECD Economic Studies*, No. 44. Paris: OECD.

Townsend, R., S. Zorya, S. Bora, and C. Delgado (2011) 'Responding to higher and more volatile world food prices', Agriculture and Rural Development Department. Washington DC: World Bank.

Walsh, J. P. (2011) 'Reconsidering the role of food prices in inflation', IMF Working Paper WP/1/71. Washington DC: IMF.

World Bank (2009) 'Rising food prices: policy options and World Bank responses', PREM, ADC, DEC. Washington DC: World Bank.

Wright, B. D. (2011) 'The economics of grain price volatility', *Applied Economic Perspectives and Policy*, 33: 32–58.

2

Food Inflation in the EU: Contrasting Experience and Recent Insights

Tim Lloyd, Steve McCorriston, and Wyn Morgan

2.1 Introduction

In recent years, there has been considerable interest from the academic and policy communities, as well as civil society, on the causes and consequences of the recent world commodity price spikes of 2007–08 and 2011 with the related concern about how best to deal with price spikes that might emerge in the future. While these events on world markets also triggered concerns about domestic food inflation, it is notable that the experience of food price changes within countries was often very different from those on world markets, particularly when we focus on price changes occurring at the retail level. For example, the International Monetary Fund noted that prices within a number of developing countries sometimes rose but in other cases remained more or less unchanged despite the spikes that were obvious on world markets (IMF, 2011). In developed countries too, the experience of price inflation differed—often quite markedly—across countries, though with the common characteristic that changes in retail prices were considerably less relative to price changes occurring in domestic and world commodity markets. This is not to underplay the importance of domestic food price inflation but to note that the behaviour of prices at the retail level is—and commonly was—very different from the characteristics of prices of commodities traded on world markets. There may be obvious reasons for these differences. First, the food products that are bought by consumers at the retail level are not the same as the 'food' products traded on world markets. Even in cases where consumers in developing countries purchased more or less unprocessed products, there would still be additional costs of packaging, transportation, and distribution to account for even more or less identical products. Second, domestic food price behaviour could differ from world prices due to the way governments responded to the commodity price spikes through the use of trade policy instruments, the provision of social safety nets, price controls, and so on which dampened down the features of domestic price changes compared to those on world markets.

While these retail price dynamics across countries are readily observed and the resulting differences in the domestic retail price behaviour from price changes occurring on world markets not so surprising, the experience of food inflation across EU Member States poses an interesting conundrum. Despite its customs union status, coupled with the supposedly integrated market associated with its single market status, exhibiting common trade, agricultural, and other EU-wide policies, the experience of food inflation across EU Member States differed considerably. Even within countries that had common monetary and exchange rate arrangements, and even if one excludes from this comparison the experience of the new Member States, the recent experience of food price inflation has varied widely across the EU. For example, coincident with the recent commodity price spike in 2011, annual retail food price inflation in Estonia reached 9.8 per cent, in Denmark it was 3.9 per cent, while in Sweden it was 1.3 per cent. In Germany, annual food inflation stood at 2.8 per cent, yet in Austria it was 4.3 per cent.

This disparity in the experience of food inflation across the EU over recent years has drawn the attention of policymakers, the focus of these concerns being that one of the main reasons for the contrasting experience in the face of shocks from world markets was the differences in the characteristics and functioning of the food chain across individual EU Member States. More specifically, the concerns related to how the food supply chain functioned, about the extent of competition throughout the supply chain, and about how price changes originating on world markets (or at least in the upstream agricultural sector) were transmitted to corresponding price changes in the retail sector. These concerns were reflected in a number of ways. Bukeviciute et al. (2009) drew the direct link between the functioning of the food supply chain and the behaviour of retail prices across Member States. This was also tied to the establishment by the European Commission of a *High Level Forum for a Better Functioning Food Supply Chain* (EU Commission, 2009), while the overall concern with competition in the food sector was reflected in the audit of competition investigations by national competition authorities across the EU (EU Competition Network, 2012).

Against this background, the overall objective of this chapter is twofold. First, we document the experience of food price inflation across the EU, highlighting the extent of the differences in the food inflationary experience and how food inflation differs from non-food inflation. This provides some context to how retail food price behaviour differs from price dynamics in other markets. Second, since the primary issue underpinning food price inflation relates to the mechanism governing the transmission of commodity shocks, we highlight which factors determine the price transmission process, and how competition in the food chain may be expected to have an impact on price transmission. We refer also to recent empirical studies of price transmission as they relate to the food sector and draw attention, where appropriate, to recent research on food inflation across the EU and how the differences in the experience of food inflation may relate to the characteristics of the food chain. Specifically, given the complexity of addressing competition issues in the food sector, we highlight specific aspects of competition and how they may affect the price transmission process.

The organization of the chapter is as follows. In Section 2.2, we assess why the food inflation issue matters in the context of the EU. In Section 2.3, we review the experience of food price inflation in EU Member States, in the process highlighting the differences between food and non-food price inflation. Given that the issues underpinning food inflation relate to price transmission, in Section 2.4 we review theoretical and empirical aspects of the price transmission process as they apply in food markets and highlight the insights that apply with respect to the differences in the structure and functioning of the food chain throughout the EU. This leads to a final issue, which is addressed in Section 2.5: given the range of factors that may determine price transmission, is it that surprising that food inflation differs throughout the EU in spite of the common policies and supposedly integrated market which is intended as a feature of the EU?

2.2 Why Does Food Inflation Matter?

If inflation is—in the words of Ken Rogoff—the most regressive of taxes (Rogoff, 2009), then food inflation is its most regressive component. For many developed countries, consumers spend a relatively small proportion of their household income on food, but the poorest spend a relatively higher share. This relationship between the share of income spent on food and the level of income is also borne out when comparing household expenditure on food across the EU Member States and is highlighted in Figure 2.1. At the aggregate level, the average expenditure on food from household income is around 14 per cent, but this average figure conceals wide differences. For example, in Romania, the share of household expenditure on food is around 36 per cent and remains well above the EU average in Lithuania, Bulgaria, Latvia, and Poland, where expenditure on food is around 23 per cent of household income. This relationship between the level of household income and the share of income would also be evident for income groups within

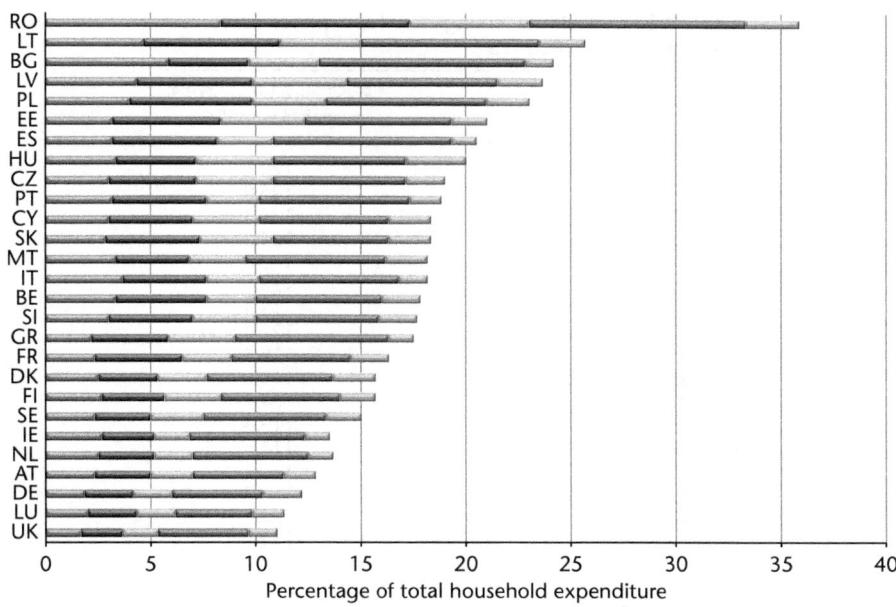

Figure 2.1. Share of household expenditure on food across EU-27
Source: Eurostat

countries: it is well known that the poor spend a greater proportion of household expenditure on food. For example, in the UK, although household expenditure on food is, on average, relatively low, at around 10 per cent of household income, the share for the lowest deciles is around twice that for the highest deciles. This is why food inflation is regressive: within any country it affects the poor most, while at the country level, food inflation has more significance depending on the overall income level of the country.

These differences across EU Member States and, by extension, within Member States highlight the significance of food inflation. We document in more detail the characteristics of food inflation in Section 2.3, but in Table 2.1 we highlight the cumulative effect of food inflation over the period 2005–13 (with 2005 = 100), the period that coincides with the world commodity price spikes. Although, as we detail in Section 2.3 food inflation is volatile, when the food price variability has been ironed out, food prices in the OECD European countries were 27 per cent higher in 2013 than at the beginning of 2005. But this average figure hides substantial differences in the cost of food. In the UK, the cumulative food price index was 39 per cent higher at the end of the eight-year period than it had been before; in Ireland, by contrast, the food price index was only 5 per cent higher in 2013 compared with 2005.

Table 2.1. Cumulative changes in food prices across EU Member States, 2005–13

Country	Food Price Index: 2013 compared with 2005
Austria	126.1
Belgium	126.0
Czech Republic	127.3
Denmark	124.3
Estonia	148.6
Finland	128.1
France	115.3
Germany	122.0
Greece	119.5
Hungary	154.7
Ireland	104.7
Italy	119.4
Netherlands	114.6
Poland	131.7
Portugal	111.8
Slovak Republic	123.9
Slovenia	133.1
Spain	122.2
Sweden	118.3
United Kingdom	138.9
OECD Europe	*127.0*

Source: OECD

Of course, these data do not indicate the peaks of food inflation and, at times, food price deflation, but they serve to indicate the considerably different experience of the cumulative effect of food inflation across the EU. Setting aside the issue of whether monetary authorities should target food inflation, the data presented in Table 2.1 indicate that for many EU Member States, the cost of food has increased since the mid-2000s and that the impact of this will be more significant on poorer countries in the EU, and among the least well-off within any Member State. In this context, note the cumulative effect of food inflation in Estonia, Hungary, Poland, and Slovenia.

Aside from the effect of food inflation on consumers, the issue about price dynamics at the retail level also raises concerns about the intensity of competition in the food chain, how price changes originating at the upstream level are transmitted to consumers, especially the way competition influences this price transmission process, and, in turn, 'equity' or 'fairness' in relations between various constituents throughout the food supply chain. Figure 2.2 relates to these concerns, where we present data on prices at either end of the supply chain for bread and wheat. There are two dimensions which are of note from these data. First, despite the dynamics of food inflation which are detailed below, prices for raw commodities (either on world markets or at the EU farm level) are much more volatile compared with the retail product.

Food Inflation in the EU

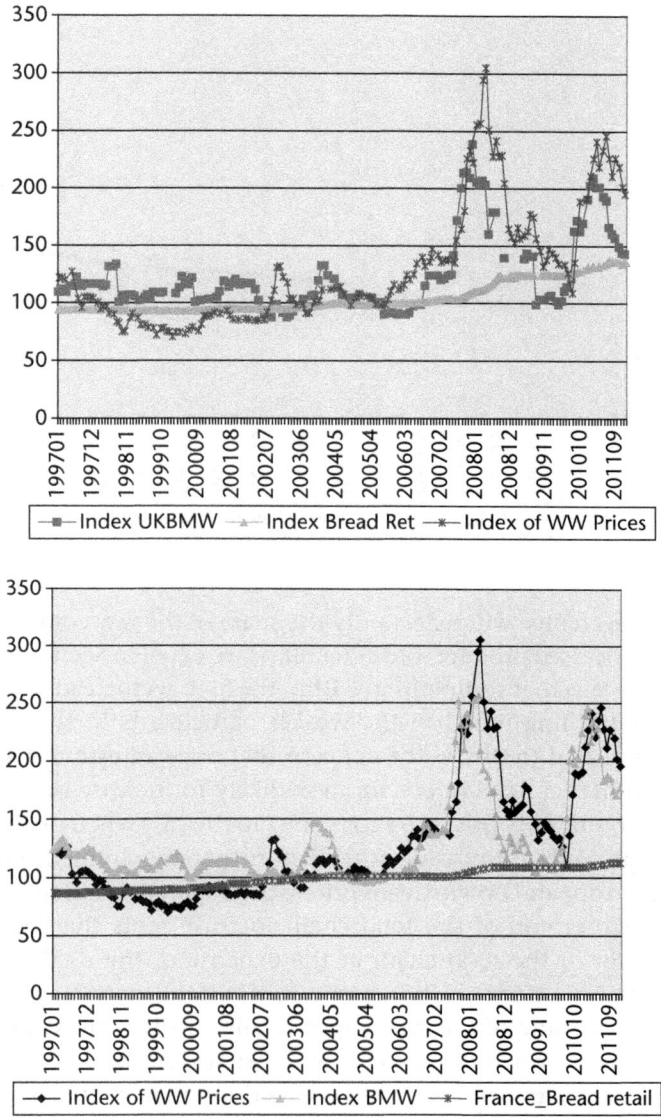

Figure 2.2. Price indices for wheat in global and domestic agricultural markets and retail bread prices, 1997–2011 for UK, France, and Poland

This implies that the retail–farm level margin is changing but that the main burden of adjustment is occurring upstream, not at the retail level. While there may be solid theoretical reasons why prices at the retail level are less volatile and are likely to change by less than a corresponding change in the raw commodity price (for example, even in competitive markets, the extent of

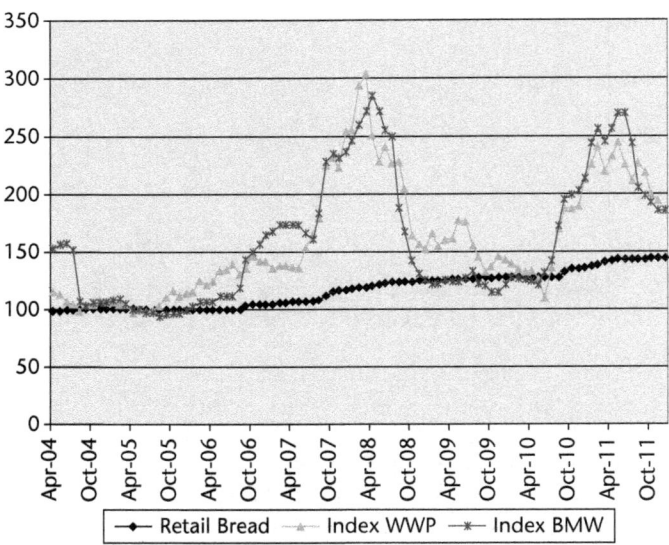

Figure 2.2. Continued

the retail price change will reflect only the share of the raw commodity input in the final processed product sold at retail), as we detail in Section 2.4, there is an overall concern that competition within the food sector leads to the burden of adjustment falling more on the weaker participants in the food sector. Second, and related to this, is the concern that price adjustment at the retail level is asymmetric: retail prices are more likely to increase by more and/or faster when commodity prices rise compared to the case when upstream prices fall. Over time, and setting aside other factors that may influence the margin between upstream and downstream prices, the apparently different dynamics of prices at either end of the food chain increase rents that accrue to the stronger parties in the food chain at the expense of the weaker parties. In large part, this is why competition authorities became increasingly concerned about the functioning of the food supply chain in the context of the recent commodity crises on world markets.[1] We take up the issue of the potential links between price transmission and competition in Section 2.4.

In sum, addressing food inflation matters for two broad reasons. First, food inflation has a direct impact on consumers, and this effect is regressive. It is also more likely to be an issue for relatively poorer Member States as this potential impact will vary across the EU. Second, retail food prices behave differently from upstream raw commodity prices, and there is a concern that

---

[1] The discussion in this chapter focuses on price transmission in food markets drawing on various contributions from McCorriston (2002) onwards though the issue also applies to other markets.

Food Inflation in the EU

the extent of competition in the food sector impacts on the price adjustment process to the detriment of consumers (not only are price rises passed on, but retail prices do not reflect declines in upstream prices) and weaker parties in the food chain who are likely to incur most of the burden of adjustment. In Section 2.3, we focus on the dynamics of retail food price inflation, and in subsequent sections we focus on the factors that may influence the price transmission process. These then form the basis for addressing whether it is surprising that the experience of food inflation varies across EU Member States.

2.3 Experience of Food Inflation across the EU

There are three principal observations to make here: (i) the experience of food inflation varies considerably across EU Member States (during both the commodity price spikes and over a longer time frame); (ii) average rates of food inflation exceed non-food inflation for most, but not all, EU Member States; and (iii) there are considerable differences in the experience with regard to the variability of food and non-food inflation across Member States. We highlight these features of food inflation below with data covering the 2000–13 period.

2.3.1 Food Inflation across the EU

In Figure 2.3, we report the experience of food price inflation across the EU. For the twenty EU Member States reported, food inflation for the 2000–13 period was, on average, 2.8 per cent, which slightly exceeds the average for the Euro Area countries of 2.5 per cent. However, these averages across the EU hide some important differences across EU Member States. Food inflation has—on average—tended to be relatively low in Ireland, the Netherlands, France, and Germany but at the higher end in the UK and Spain. Food inflation has been notably higher in the new Member States, with average annualized monthly inflation of 6.4 per cent in Hungary, 4.7 per cent in Estonia, and 3.6 per cent in Poland.

Figure 2.3 also highlights the contrasting experience of food inflation that coincided with the commodity price spikes on world markets in 2007–08 and 2011. The (average) differences are more marked in the sub-period 2007–08 and 2011 with, for the selected Member States, notable differences between the UK and Sweden. These averages hide some important peaks in food inflation which, in the case of the UK, reached around 12 per cent.[2] Ireland also stands out; food inflation was actually lower during the commodity price spike periods.

---

[2] For other EU Member States, food inflation peaked at even higher rates: most notably, Bulgaria (23 per cent), Latvia (20 per cent), Estonia (17 per cent), and Slovenia (13 per cent).

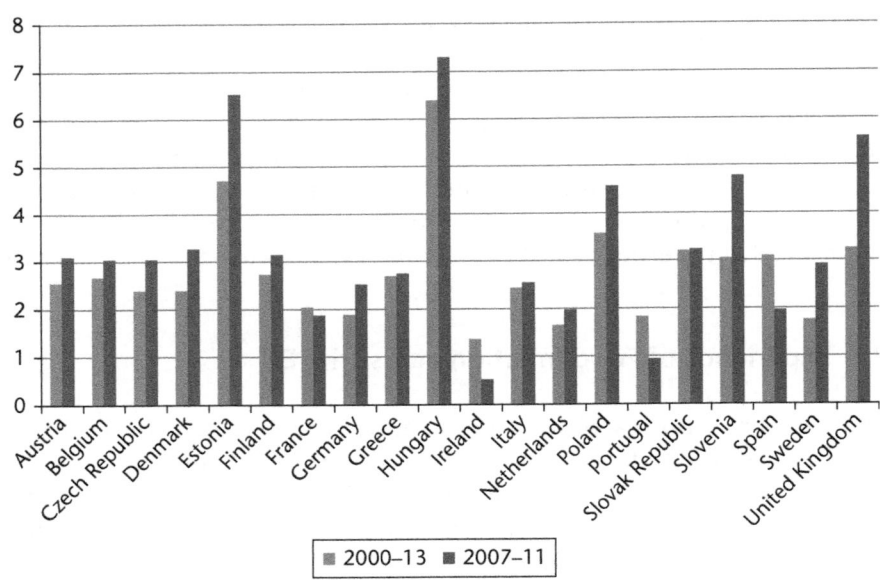

Figure 2.3. Average rates of food inflation across EU Member States, 2000–13

2.3.2 Food versus Non-Food Inflation

Figure 2.4 highlights the differences between food and non-food inflation across the EU. For the EU as a whole, food inflation (on average) considerably exceeds non-food inflation, though the extent of the difference between these two sources of inflation is less for Euro Area Member States. Food inflation for the most part exceeds non-food inflation, the most significant differences being in the UK and Sweden, where food inflation has been on average more than twice that of non-food inflation. For the new Member States (particularly Estonia and Hungary), the difference between food and non-food inflation is notable. Ireland again stands out in that the average rate of non-food inflation exceeded food inflation (which is also the experience of Portugal and the Netherlands).

2.3.3 Variability of Food and Non-Food Inflation

The potential concerns of high food inflation are compounded by its variability; despite the impact of food inflation on the cost of living, the high variability of food inflation makes it more difficult for monetary authorities to address it without the risk of exacerbating output variability. The issues associated with targeting food inflation, despite the high levels that have been

Food Inflation in the EU

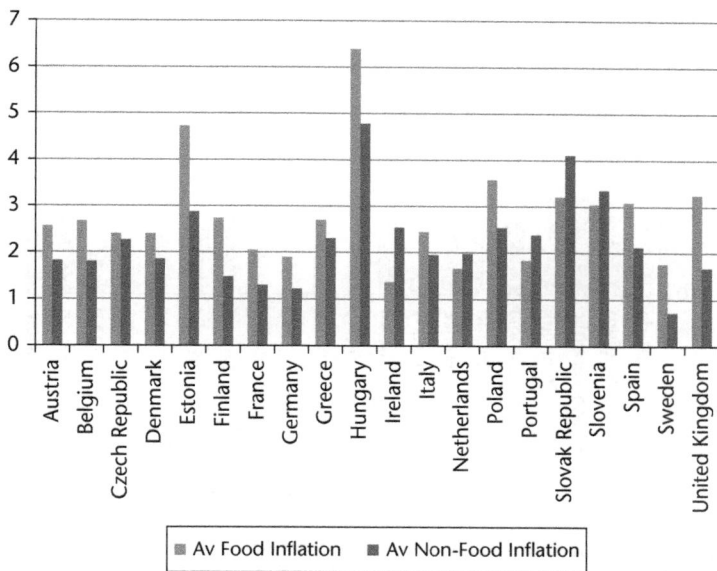

Figure 2.4. Average rates of food and non-food inflation across EU Member States, 2000–13
Source: OECD

witnessed in recent years, have been summarized in IMF (2011) and Walsh (2011), with issues associated with persistence or second-round effects being addressed by Cecchetti and Moessner (2008). The variability of food inflation across the EU is highlighted in Figure 2.5, which also presents evidence on the variability of non-food inflation by way of comparison. For the most part, the variability of food inflation compared with non-food inflation is more significant than the comparison with the levels of inflation presented in Figure 2.2; for the twenty EU Member States reported in Figure 2.4, the variability of food inflation is more than twice that of non-food inflation. Only in Sweden is the variability of non-food inflation higher than that of food inflation.

In sum, the experience of food inflation across EU Member States varies considerably, in terms of levels (these differences being more exacerbated during the recent world commodity price spikes), in terms of the differences with non-food inflation, and also in its variability. The new Member States stand out as having the highest and most variable levels of food inflation, but even within the more established members of the EU there are notable differences. Given that the transmission process is key to addressing the links between retail and world price changes, the factors that determine price transmission are addressed in the following section.

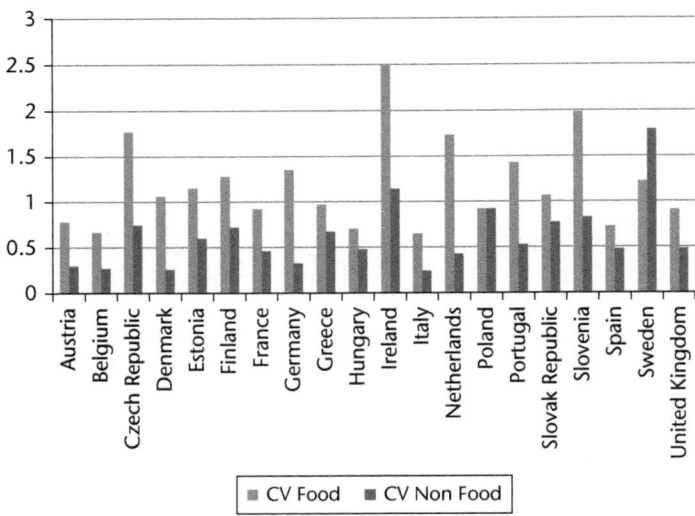

Figure 2.5. Coefficient of variation of food and non-food inflation across EU Member States, 2000–13
Source: OECD

2.4 Insights from Price Transmission

Key to understanding the behaviour of retail prices and the dynamics of food inflation, particularly in the face of shocks emanating from world markets, is the price transmission process. It is this process which underpins a wide range of studies of inflation more generally (for example, in exploring the impact of oil price shocks on inflation) through to more micro-orientated studies that aim to focus on price transmission in specific sectors. While, in principle, benchmarking the expected effect of upstream shocks should be straightforward, there is considerable interest in how competition affects the price transmission process. In this section we review the literature on price transmission, highlighting both theory and insights from recent empirical studies. This forms the basis for Section 2.5: if the characteristics and the nature of competition affect the behaviour of retail prices following commodity price shocks, given the notable differences that exist in the structure and degree of competition in the food sector across EU Member States, is it any surprise that the experience of food inflation has differed so much across the EU?

In this section, we explore the links between competition in the food supply chain and how this has an impact on prices being passed from one end of the food chain through to the retail sector. We start with a basic framework to highlight the key issues, then comment on more recent developments that relate directly to food supply chain issues. Prior to focusing on food sector-specific issues, it should be noted that the issue of price transmission crosses with other areas of economics. For example, the issues addressed below have close ties with developments in macroeconomics (Blanchard and Gali, 2007) and international economics, in particular exchange rate pass-through (see, e.g., papers by Dornbusch (1987) and Feenstra (1989) and, more recently, Goldberg and Hellerstein (2008)). Public economists have been concerned with the incidence of taxes (Anderson et al., 2001), and industrial organization economists with how market structure matters for the incidence of cost changes (see, e.g., Dixit, 1986). Although the context sometimes differs, there are common mechanisms that determine how cost, tax, or exchange rate changes pass through to consumer prices and the role that competition plays in determining the transmission process. But, as we discuss in more detail in Section 2.4, there are food sector-specific issues that should also be accounted for in this assessment.

It should be noted that there is already a large empirical literature on the price transmission process in agricultural and food markets. Vavra and Goodwin (2005) explore these issues with respect to price transmission in the food chain. Often motivated by widening margins between agricultural prices and downstream (usually consumer) prices, this strand of the literature has employed time series data to estimate the extent, rate, and nature of the price adjustment process. This approach is informative in outlining the general features of the functioning of the food supply chain, with the econometric approach being flexible enough to address several interesting issues with respect to food price adjustment. The methodology can also be used to address asymmetric price adjustment as well as non-linear aspects of price behaviour; for example, that small cost changes may not be passed through to consumers but large price changes are.

While informative and, subject to data availability, relatively straightforward to apply, one problem with this approach is that it is largely *a*theoretical. In this context, it becomes difficult to 'explain' any of the results that are produced. For example, the econometric results may indicate imperfect price transmission; but since there can be many factors that influence pricing in the food supply sector, it becomes difficult to ascertain what has caused this observed lack of pass-through. Similarly, while the econometrician may conjecture that the lack of competition may result in asymmetric price transmission, in the absence of any structure to the underlying framework it is difficult to 'blame' imperfect competition for this observation. At best, time series approaches (which are typically conducted using pairs of price series) represent a 'first-pass' test relating to the functioning of the food chain since it is difficult to ascertain from this framework what factors are likely to be driving the results.

With this in mind, we focus in this section on the links between competition and the price transmission process from a theoretical perspective. We start by benchmarking the case with perfect competition and then add in imperfect competition to a food sector with a single stage. We then build on this to highlight how other factors may influence the overall outcome, including the role of buyer power. This is followed by more recent research—both theoretical and empirical—that addresses these issues.

2.4.1 Competition and Pass-Through: Basic Insights

The main intuition relating commodity price changes to competition can be highlighted by drawing on the model of farm–retail price changes initially developed by Gardner (1975). In the Gardner set-up, there is a 'food sector' that combines agricultural inputs and marketing services inputs (e.g., other inputs, packaging) which are procured and subsequently sold by the food sector; there is no distinction here between retail and processing as different activities. Combining these alternative sources of inputs is dependent on the nature of technology, either variable or fixed proportions, the main issue here being the extent of substitutability between the agricultural and non-agricultural inputs; with a fixed proportions technology, the elasticity of substitution between these two sources of inputs is zero. Aside from the technology issue, the extent of price transmission is determined by the elasticities of supply in the two upstream (agriculture and non-agriculture) markets and the demand elasticity at the retail level. Also in this set-up, the extent of price transmission may depend on the source of the shock, whether this occurs at the demand side or in one of the upstream markets.

This model was adapted by McCorriston et al. (1998) to account for imperfect competition. In this framework, there is a single intermediary stage that is labelled the 'food processing/retailing sector' that produces a homogeneous good with firms pursuing quantity-setting strategies. At an intuitive level, McCorriston et al. show that the extent of price transmission arising from the impact of shocks occurring at the farm stage on retail prices can be separated into two parts:

$$PT = f(\Delta MU, \Delta C) \qquad (1)$$

that is, the extent of price transmission (PT) depends on the change in the aggregate markup (MU) for firms that constitute this intermediate, oligopolistic food sector and the change in costs (C). Assume initially that the food sector is competitive such that the markup is zero. The extent of price transmission will then depend only on the change in costs. If a fixed proportions technology is assumed, then the extent of price transmission will reflect the share of agricultural raw materials in the competitive food industry cost function. So, if the share of agricultural raw materials in the food industry costs function is 50 per cent, the transmission elasticity should be 0.5.[3] This insight from a competitive market is embedded in the Gardner (1975) framework noted above.

However, if the markup is positive indicating a departure from the competitive assumption, then market power influences the degree of price transmission. Furthermore, the markup may alter in response to a change in food industry costs, the extent of this change depending upon the degree of market power (itself contingent on the nature of competitive interaction between firms and the number of competing firms) and the nature of the demand function. Interestingly, the analysis shows that if the demand function is log-linear (i.e., constant elasticity), the markup remains constant in the face of cost shocks and hence does not influence the degree of price transmission when costs change. However, in other circumstances, the change in the markup plays a role when costs change: specifically, the markup falls when costs rise and therefore serves to reduce the price transmission elasticity. In this case, there is 'under-shifting' and retail prices will change less than farm-gate prices.

To see the issues more directly, McCorriston et al. (1998) derive a price transmission elasticity involving an upstream agricultural market with an imperfectly competitive food industry which has as its arguments the characteristics of the agricultural and non-agricultural markets, the nature of technology between these two markets, and the retail demand function (as in the basic Gardner model) as well as accounting for the change in the (non-zero) markup in the food sector. This price transmission elasticity is given by:

$$\tau = \frac{S_A(1+\gamma\sigma)}{(1+S_A\gamma\sigma)(1+\mu)+(S_B\gamma\eta)} \qquad (2)$$

where S_A is the share of the raw agricultural commodity in the food industry cost function, S_B is the share of other inputs, σ is the elasticity of substitution between agricultural and materials inputs, γ is the inverse elasticity of supply of marketing inputs, and η is the industry elasticity of demand. The effect of competition enters via the μ parameter, which relates to the elasticity of the industry markup given by $\mu = \omega(\theta/n\eta - \theta)$ with ω representing the change in the elasticity of demand for a given change in the retail price, where n is the number of competing firms and θ is a measure of the intensity of competition between firms.

---

[3] Even with a competitive food industry, there may be imperfect price transmission if we have a variable proportions technology as in Gardner (1975). However, the role of the substitution elasticity is likely to be swamped even by relatively low degrees of market power as shown in McCorriston et al. (1998), so for simplicity, we will confine the discussion to a fixed proportions technology.

There are several observations that can be made with respect to this price transmission elasticity. First, and most obviously with reference to the discussion made above about time series econometric studies, the influences on price transmission are manifold. Market power may matter, but it may not be the only factor that has an impact on the relationship.

Second, it is useful to isolate the specific role market power may play in determining price transmission. To see this, assume a zero value for the elasticity of substitution between agricultural and other inputs together with a perfectly elastic supply for marketing inputs, that is, σ = γ = 0. Recall that when there is no market power in the food sector (for example, n is sufficiently large that the market is competitive), then the price transmission elasticity will be given by:

$$\tau_c = S_A \tag{3}$$

In other words, the change in retail prices should equal the share of agricultural inputs in the food industry cost function; this should be the extent to which, if markets are competitive, retail prices change given the corresponding change in input prices.

The effect of market power on price transmission in the food sector depends on how the food industry markup changes, which is itself determined by $\mu = \omega(\theta/n\eta - \theta)$. This is the key point about competition and pass-through. It is not just about the number of firms and the intensity of competition; it is how these interact and thereby influence the change in the markup. In this context, even if we have a highly concentrated food sector (e.g., n = 2) and competition between these firms is not 'too' intense, the change in the markup will also depend on the nature (or, more formally, the convexity) of the demand function (ω). If we had a log-linear demand function (meaning that ω = 0), for example, it is easy to see that the change in the markup would be zero no matter what the structural characteristics of the industry actually were. In terms of the structural characteristics of the food industry, competition need not be restricted to the number of competing firms, but could also include the competitive behaviour of these firms.

With these comments in mind, to see how imperfect competition influences the price transmission outcome, employing the same assumptions as above (σ = γ = 0) the pass-through effect can be given by:

$$\tau = \frac{S_A}{(1+\mu)} \tag{4}$$

As $\mu > 0$, then imperfect competition in the food sector dampens the price transmission effect. To see this in a different way, comparing this elasticity with the pass-through elasticity in a competitive food sector, then we have:

$$\frac{\tau_c}{\tau} = 1 + \mu \tag{5}$$

With a linear demand function (so that ω = 0), less than intense competition between firms (θ > 0) and n sufficiently small, then μ > 0; intuitively, the markup in the downstream food sector falls in the face of the increase in costs relative to the competitive baseline. In sum, subject to conditions on the demand function, market power in the food sector will lead to under-shifting of retail food prices. So, if the agricultural input accounts for 25 per cent of the food industry costs, the transmission elasticity will be less than 25 per cent.

The main insight from the above is that with a change in the costs purchased by the food sector, there are essentially two factors which will determine how food prices will change. The first is the share of costs in the industry cost function. If the food industry is competitive, this will be the only factor which will matter. However, if the food industry is imperfectly competitive, the effect on food prices will depend on how the food industry markup changes. Conditional on the assumptions on the demand curve, the industry markup will fall and retail prices will rise by less than the increase in costs. In other words, the imperfectly competitive food industry absorbs some of the cost increases and market power in the food sector serves to dampen price transmission, contingent on the characterization of the retail demand function.

What other characteristics of the food industry will be likely to matter in determining this pass-through effect? One potential factor is economies of scale; even if there was evidence of a high degree of market power, this may be offset by efficiency effects (see Morrison-Paul (2001) and Bhuyan and Lopez (1997)). Millàn (1999) has also documented the existence of economies of scale in the Spanish food sector.

McCorriston et al. (2001) have developed the framework outlined above to derive the price transmission elasticity to account for scale effects. The scale effect is captured by a parameter ρ; with ρ greater (equal, less) than 1, this represents increasing (constant, decreasing) returns to scale. McCorriston et al. (2001) amend the price transmission elasticity to account for this feature of the food industry cost function, the comparison with the competitive benchmark (subject to assumptions made about other parameters) now being given by:

$$\frac{\tau_c}{\tau} = 1 + \mu - \eta(\rho - 1)/\rho \tag{6}$$

If we have constant returns to scale (ρ = 1), then we retrieve the comparison between the competitive and imperfectly competitive case as noted above. With increasing returns to scale and assuming the retail demand function to be not 'too' convex, the under-shifting effect now weakens; conditional on the extent of the scale effect, it could be the case that food prices rise by more than the cost increase such that we would have 'over-shifting' rather than under-shifting.

Another characteristic of the food industry that draws attention from stakeholders in the food chain and policymakers is the existence of buyer power. This too can affect the transmission elasticity, as has been explored by Weldegebriel (2004). He shows that the existence of oligopsony power may offset the effect of oligopoly power in determining the effect of cost changes on food prices. Specifically, with (seller) market power in the food sector, the change in the markup determines what the price transmission elasticity will be (as we have noted above): when buyer power exists, what is important is the change in the markdown. The change in the markdown is dependent upon the extent of competition in the procurement market and the functional form of the supply function. If the markdown increases, then this increases the price transmission elasticity and offsets the reduction in the markup. If oligopsony and oligopoly co-exist, it will be difficult to ascertain what aspect of market structure and competition is determining the (net) price transmission effect.

2.4.2 Extensions

The above insights have the advantage of using a common framework to identify how competition in the food sector impacts on price transmission following a change in agricultural prices. We depart from this common framework to consider some extensions that are pertinent to the characterization of the food sector.

Take, first of all, the observation that the food supply chain is a complex series of inter-related markets that could be characterized by imperfect competition at each stage and return to the characterization that we are focusing on: seller power only. In this case, we would have successive oligopoly. The issue then is how cost changes from the agricultural sector are passed through this chain of imperfectly competitive markets. McCorriston and Sheldon (1996) show that as the number of stages in this vertical chain increases, price transmission decreases below that expected in the single stage case. However, the extent of the decline is not a simple multiple of the single stage case, since in their framework the perceived derived demand function facing each stage is contingent not just on the degree of market power at that stage (i.e., horizontal market power), but also on the degree of market power at succeeding stages. With this mechanism, market power throughout the successively oligopolistic food chain exacerbates the degree of under-shifting.

Assuming linear demand, the insight is straightforward since what determines the final change in food prices is the change in the markups at each stage in the food chain. Even if we assume arm's length pricing between each of the stages in the food chain (i.e., the downstream firms take the upstream price as given and there are no non-linear pricing contracts characterizing the links between stages in the food chain), market power at each successive stage determines what the change in the final price will be. With this characterization of the food supply chain, we have markups due to market power at both the retail and food processing sectors, the combination of these markups being known as 'double marginalization'; the existence of these 'double' markups means that the inefficiency of the food supply sector is exacerbated by the existence of market power at each stage.

As above, assume that the agricultural price (the cost to the food manufacturing sector) increases. What determines the markup in this intermediate sector is not just the extent of competition at that stage but the nature of the derived demand function facing that stage, which itself depends on the extent of competition at the retail level. The extent to which this cost is passed through to retail (assuming linear demand) will depend on how the food manufacturing industry markup changes. Markups in the retail sector are now not only determined by the intensity of competition at the retail stage but also by the level of costs arising from the intermediate stage. As costs are passed through the food manufacturing stage (albeit diluting the initial agricultural cost increase), then the change in the final retail price will be determined by the extent to which retail markups change. Taken together, and conditional on the assumptions concerning the demand function, the vertically related nature of the food chain exacerbates the extent of under-shifting that is likely to arise from imperfect competition.

Note that in the case of successive oligopoly outlined here, we have assumed arm's length pricing, but how we characterize the links between the vertical stages is also an important feature of the food supply chain. In the McCorriston and Sheldon (1996) study, the degree of under-shifting is exacerbated due to the existence of double marginalization. Any contract between, say, the food manufacturer and retailer that diminishes the double marginalization effect should have an effect on pass-through. For example, if the contract between retailers and manufacturers had the equivalent effect of vertical integration, price transmission would increase (at least relative to the successive oligopoly/arm's length pricing case).

Recognizing the chain aspect of the food supply sector raises further questions regarding the transparency of food prices in the various stages. First, measuring the effect of alternative vertical contracts on the price transmission is likely to be problematic. Second, and perhaps more practically in terms of empirical research, we need to know not just how agricultural and retail prices change but also how intermediate prices respond. This poses a challenge for recent research addressing the price transparency issue, with the focus on wholesale and retail prices thus excluding price changes from further upstream. We return to these issues in Section 2.4.

A further aspect of the framework presented above is that it assumes that the firms in the food sector are symmetric—that is, that the market is split evenly between them, implying they have equal cost structures and market shares. This simplifies the theoretical framework considerably, though it does not settle easily with the characterization of the food sector outlined above. Does it matter? Whether the focus is at food manufacturing or retailing, it is obvious that firms are not of equal size, having identical market shares and the same costs. Rather, costs vary and some firms are market leaders while others have much smaller market shares. In this case, we would have to amend the basic price transmission story outlined above: the aggregate markup will change not just because of the change in costs, but also because the change in costs affects each firm to varying degrees. As such, the level of concentration may change as the fall in costs favours larger, lower-cost firms more than smaller, higher-cost firms. Dung (1993), for example, shows that market power will increase with rising levels of concentration. In turn, in the context of the above outline, this will serve to lower the degree of price transmission.

2.4.3 *Matching Empirics with Theory*

At the start of this section, we made reference to the time series studies that explore the nature of price transmission from raw agricultural prices through to prices at the retail end. Since they are not reliant on a specific theoretical framework, it becomes difficult to fully reflect on the outcome as the econometric model does not make it possible to assess what drives a specific result that is produced. So, while a less than one-for-one price transmission effect may be consistent with the presence of market power in the food sector, we cannot be sure that this is the only factor that is driving this result. More desirable are results and insights produced on the basis of structural models where the demand function is appropriately estimated and market power allowed for. Thus, while time series econometrics are informative and are often parsimonious in terms of data requirements, the insights they offer in terms of how competition affects price transmission can be quite limited.

There have been some attempts at estimating structural models; while these are not confined to food markets, they do give some insights. For example, Barnett et al. (1995) estimated a structural model for the US tobacco industry. They showed that in the presence of market power, taxes are over-shifted. However, it should be noted that their model allows for increasing returns to scale which, as discussed above, may outweigh the influence of (under-shifting caused by) imperfect competition between firms. Delipalla and O'Donnell (2001) have estimated the incidence of taxes in the European cigarette industry and found evidence of under-shifting of taxes among the largest EU countries, though some evidence of over-shifting arises in other countries. In terms of the European food sector, Bettendorf and Verboven (2000) have estimated a model of the Dutch coffee industry and found evidence of under-shifting of raw coffee bean prices on retail prices.

Kim and Cotterill (2008) were among the first to estimate a structural model which allowed for product differentiation between brands with an application to the food sector. With the data relating to the US processed cheese market, the change in costs will refer to the price of raw milk. Estimating a discrete choice model that allows them to estimate (own and cross) price elasticities at the brand level, they show that price transmission will depend on the substitutability between brands. They simulate pass-through for two alternative characterizations of firm behaviour, one where the market is fairly competitive (Bertrand-Nash pricing) and the other where pricing is collusive. In the case of competitive pricing, pass-through of cost changes is almost complete; but with collusive pricing, there is a considerable reduction in pass-through. Since the estimates are at the brand level, the extent of pass-through also varies by brand. In aggregate, the transmission elasticity for collusive pricing is estimated to be around 85 per cent lower than the competitive case.

2.4.4 Recent Developments

Research has developed on the issue of linking price developments with competition in food markets in recent years. We discuss briefly some of these more recent contributions and highlight the insights they bring.

2.4.4.1 DECOMPOSING PASS-THROUGH

An important contribution to understanding the links between price transmission and competition in the food sector comes from the work of Nakamura and Zerom (2010). They focus on the US coffee sector largely due to the availability of data and the ability to trace the raw agricultural input (raw coffee beans) through to the retail stage. Also, given the nature of the coffee supply chain, they can easily allocate the share in costs to the raw commodity input and other costs. They make several contributions to the issue. First, they allow for sticky price adjustment in the form of menu costs. This ties with research in macroeconomics on the micro-foundations of inflation which suggests that prices may be sticky in the sense that, due to menu costs (i.e., changing prices is costly), firms adjust prices infrequently. Second, they estimate a structural model that allows for product differentiation at the retail stage. Third, they can retrieve a measure of the markups and assess how they change in face of cost shocks to the price of coffee.

The results are insightful. Overall, long-run pass-through is relatively low in the coffee sector, with a 1 per cent shock in costs leading to around a 0.3 per cent increase in retail prices. There are several factors that lie behind this result. Specifically, despite menu costs being relatively low, they are shown to contribute to short-run price stickiness and delayed response to shocks. More importantly, the low pass-through arises because downstream firms reduce their markups by around one-third. As we have noted in the outline above, part of the mechanism of this effect is due to the change in the price elasticity of demand that depends on the curvature of the demand function. Borrowing the terminology from Klenow and Willis (2006), they refer to this as 'super-elasticity' (the percentage change in the price elasticity for a given percentage increase in prices) which they estimate to be relatively high at 4.6 per cent; it is the nature of this change in the price elasticity due to the shape of the demand function which gives rise to a substantial change in firms' markups. While not the only change that determines the overall price transmission effect, it is nevertheless indicative that how firms' markups change is an important aspect of the price transmission effect and confirms the importance of this specific mechanism, which we highlighted in Section 2.4.1.

2.4.4.2 ASYMMETRIC PRICE ADJUSTMENT

One of the common insights from time series econometric studies of price transmission is the existence of asymmetric price adjustment; this ties in with common concerns about price adjustment in the food sector and elsewhere, that is, that the food industry is quick to pass cost increases on to consumers but less willing to reduce prices when costs subsequently decline. Meyer and von Cramon-Taubadel (2004) provide a review of these issues in agricultural and food markets. Peltzman (2000) explores whether this issue could be tied to concentration, though insights from theoretical models on this issue have been limited. However, Tappata (2009) shows that concerns about competition need not be the cause of asymmetric price transmission but that asymmetric price transmission can arise in competitive markets with partially informed consumers.

Richards et al. (2012) have explored this issue in the context of the recent experience with food price inflation. They show that the pricing conduct of firms varies with the direction of underlying commodity price shocks but that the outcome can vary by commodity sector. For example, for one commodity (potatoes), when commodity prices are rising, the industry markup decreases; but when prices subsequently fall, the markup widens. They also show that the increase in the markup in the declining price phase is greater than the reduction in the margin when commodity prices are rising. In the other commodity sector they explore (fluid milk), these asymmetric effects do not exist, though there is still some degree of asymmetry to the extent that when commodity prices are falling, margins are unchanged; but when they are rising, margins narrow.

2.4.4.3 MULTI-PRODUCT RETAILERS

Most, if not all, of the research that forms the basis of the previous discussion relates to single product firms; with imperfect competition, under-shifting is likely to arise unless the demand function is sufficiently convex. Yet, food retailers are multi-product outlets selling a wide variety of products and competing along a wider range of attributes. Previous work allowing for product differentiation did not tie down specific outcomes where product differentiation reversed the outcomes or insights significantly from what we have noted above. For example, in Anderson et al. (2001), the convexity of the demand function still plays a crucial role in determining whether over- or under-shifting will arise.

Hamilton (2009) has made an important contribution to understanding the links between the extent of price transmission and the existence of multi-product retailers. In essence, there are two aspects at play: the first is the change in costs for a particular product; the second is the number of varieties put on sale by the multi-product retailer. Hamilton shows that as costs increase, the retailer puts fewer varieties on sale. This softens price competition such that the net effect on the retail price of the good is larger than the initial increase in costs. Though we have noted above that over-shifting of cost increases could arise with the demand function being sufficiently convex, under-shifting is more likely. However, in this case, over-shifting does not depend on the curvature of the demand function. More specifically, if the demand function is concave, over-shifting is likely to arise. The over-shifting effect among multi-product retailers arises because variety withdrawal weakens the extent of competition in the retail market.

Hamilton and Richards (2011) have explored this issue empirically using detailed retail price data from the US ready-to-eat cereal market. They show that, in isolating the pass-through effect without the variety effect, pass-through of costs is indeed less than one (our expected outcome). But when accounting for the variety withdrawal effect due to the increase in costs, pass-through increases above the initial increase in costs. These results support the idea that accounting for the multi-product nature of supermarket retailers can give an outcome that would not arise in the standard framework and show that this particular characterization of the food sector has to be accounted for in gauging the overall effect.

2.4.4.4 PASS-THROUGH AND VERTICAL RESTRAINTS

We have noted above that in the context of food supply chains, market power at each stage can have an impact on the overall price transmission effect

(McCorriston and Sheldon, 1996). We noted that the common assumption was arm's length pricing and the non-existence of non-linear pricing contracts in characterizing the links between the food processing and retailing sectors in the food chain. However, in reality, vertical restraints, which can take a variety of forms, are an important aspect of the relation between retailers and manufacturers. While McCorriston and Sheldon assume arm's length pricing—and therefore that the extent of double-marginalization determines the cost pass-through in the successive stages in the vertical chain—it may nevertheless be likely that vertical restraints have an impact on the price transmission outcome. Intuitively, if double marginalization is the 'benchmark outcome', to the extent that vertical restraints ameliorate this effect they will also affect the extent of pass-through.

This issue has been explored recently by Bonnet et al. (2010). Estimating a structural model using data from the German coffee market, they explore how non-linear pricing and vertical restraints such as wholesale price discrimination affect the pass-through of costs from the upstream sector. Benchmarked against the linear contract (arm's length pricing), they show that the existence of vertical restraints serves to increase price transmission. With a 10 per cent increase in costs, the linear pricing case leads to an average 7.2 per cent increase in retail prices. The main result from their paper is that vertical restraints in the form of resale price maintenance increase the level of pass-through. The intuition they offer is that, in the presence of resale price maintenance, when there is a cost shock, the existence of the vertical restraint limits the ability of firms to adjust their markups. Since we know from previous discussion that the change in the markup can help to reduce the price transmission effect, since the vertical restraint limits the extent to which firms can adjust their markups, pass-through therefore increases.

Bonnet and Réquillart (2012) apply a similar framework to the EU sugar sector. Again, they allow for the existence of vertical restraints between soft drink manufacturers and the sugar processors, though they do not explore the range of outcomes with alternative characterizations of vertical restraints. However, they do highlight the pass-through effect at the brand level and draw a distinction between national brands and private labels. In aggregate, there is over-shifting (retail prices change by more than the change in costs), but the transmission effect varies by between 1.1 per cent and 1.23 per cent at the brand level, that is, the effect of a 10 per cent cost increase can increase the price of certain brands by 12.3 per cent. On the whole, the pass-through effect is greater for national brands than private labels, though this may be due to differences in the initial markups between branded and private labels.

2.4.4.5 SUMMARY

In large part, concerns about the functioning of the food sector relate to transparency in the pricing of food products; in the context of recent developments in world and domestic agricultural markets coupled with high levels of food inflation across many OECD countries, there is concern that the competitive aspects of the food sector can affect the extent of price transmission. We next address whether the differences in market structure across the EU can provide insights into the experience of food inflation.

2.5 Is it Surprising that the Experience of Food Inflation Varies across the EU?

This chapter was predicated on the observation that the experience of food inflation has varied considerably across EU Member States and that the concerns it raises relate not just to the effect on consumers, but also to more general concerns about how the food chain functions. In the face of the commodity price shocks that have been witnessed on world markets in recent years, the key mechanism in understanding how prices respond at the retail level is the price transmission process. In practice, there are many factors that can influence this effect: the extent to which countries are exposed to world markets may influence the magnitude of this shock; since commodities on world markets are priced in US dollars, exchange rate movements may exacerbate or dampen the effects of these shocks. But recent attention in policy circles has focused on competition in the food sector as one of the main factors that can have a bearing on this process. This is borne out by the theoretical and empirical literature on price transmission: the characteristics of the food sector potentially matter in determining the price transmission effect, though the ways in which they do so are complex. However, taking the basic principle that the structure and extent of competition in the food chain matters, and given that differences in the food sector across EU Member States are likely to exist, is it surprising that the experience of food inflation differs?

Addressing competition in the food sector is complex: the issue is not just about firm numbers as reflected in concentration ratios but about behaviour, about where market power may be exercised (i.e., selling or buyer power or both), the form of market power (i.e., vertical restraints or other practices that influence relations between parties in the food sector), whether the growth of private labels have a pro- or anti-competitive effect, and so on. These aspects of the food sector, together with the multi-product nature of food retailing, pose immense challenges in addressing these issues even in a national context. How the dimensions of the food sector can contribute to explaining the differences in the experience of food inflation across the EU Member States is even more of a challenge given the (more general) difficulty in accessing consistent data on the food sector rather than just retail and product prices.

Specifically, the characterization of the food chain varies across EU Member States. In some cases retail concentration is high (e.g., the UK), while in others it is comparatively low (e.g., Italy). In some cases the role of hard discounters has been an important feature of the competitive environment in the food sector in recent years (e.g., Germany), while in still others the penetration of discounters has been weaker. In some Member States, the penetration of private labels has been limited, but it has increased rapidly in others. The process of consolidation in the food sector has been obvious in some Member States, but due to differences in regulation and the nature of corporate governance as well as mergers and acquisitions, the market for corporate control is thinner in other Member States, which therefore forego some of the positive effects (e.g., increases in efficiency) yet avoid the potential negative effects (e.g., impacts on competition) that may accompany industry restructuring. Combined with the lack of consistent time series data detailing these aspects of competition in the food sector across the EU, providing a conclusive picture of competition in national food chains across EU Member States is currently an insurmountable task.

Recent research on food inflation across the EU by Lloyd et al. (2014) provides some insights linking the varied experience of food inflation in the EU to differences in the food sector. They proceed in two steps. First, they specify and estimate a structural vector autoregressive model of retail food inflation focusing on a specific commodity chain (bread and wheat) across eleven EU Member States (Austria, Belgium, Denmark, France, Germany, Italy, the Netherlands, Portugal, Spain, Sweden, and the UK). The structure of this model is common to all eleven Member States, with retail bread prices being linked to factors such as exchange rates and macroeconomic conditions in each of the Member States, together with world wheat prices and world oil prices. Applying this framework, they derive country-specific long-run price transmission elasticities and then derive the extent to which commodity price shocks originating on the world market contribute to the behaviour of retail bread prices in each of the Member States.

The long-run price transmission elasticities are reported in Table 2.2. The results highlight considerable differences in these long-run elasticities. With the average for the eleven countries being 0.31, the transmission effects are notably higher in Spain, Sweden, and Italy and considerably lower than the average in the Netherlands and France. The significance of these elasticities relates to the fact that the mechanisms involved in the food inflation experience are likely to differ across EU Member States.

Their analysis also sheds light on the relative importance of factors such as exchange rate, domestic demand, and the prices of wheat and oil on the

Table 2.2. Long-run price transmission elasticities: Retail bread prices w.r.t. world wheat prices

Country	LR Price Transmission
Austria	0.25
Belgium	0.22
Denmark	0.39
Germany	0.23
France	0.08
Italy	0.45
Netherlands	0.14
Portugal	0.27
Spain	0.49
Sweden	0.43
UK	0.33
Average	0.31

evolution of retail bread prices. Their results suggest that for most countries (particularly Austria and Germany), developments in the food sector itself are more important to domestic bread prices than developments in commodity prices. As an average of the eleven countries they analyse, 44 per cent of the variation in retail bread prices are accounted for by developments in domestic retail sectors. For comparison, wheat and oil account for 39 per cent and 10 per cent respectively. Given that we know that the experience of food inflation has varied across EU Member States, that the price transmission mechanism varies considerably, and that the contribution of world price shocks also differs, the next question relates to why we observe these differences and how these differences may be linked to the characteristics of the food sector in each EU Member State.

Using estimates of the contribution of world wheat prices in retail bread prices, Lloyd et al. (2014) correlate the experience of food inflation with measures relating to (or at least proxy for) the functioning of the food sector in these countries. Given the lack of consistent and comparative data detailing competition in the food sector across the EU, they are confined to correlating the cross-country experience of price transmission with broad cross-country aggregates characterizing competition in the food sector for the EU Member States they cover. Two examples of the insights linking food inflation with features of the food sector are illustrated in Figures 2.6 and 2.7.

In Figure 2.6, they take the measure of the contribution of changes in world wheat prices and correlate this with a measure of 'barriers to competition' in the retail sector in the EU. This measure of 'barriers to competition' (high values represent increased barriers) is not a direct measure of the intensity of competition between established retailers, nor of the extent to which vertical linkages in the food chain reflect the exercise of market power, but is rather a

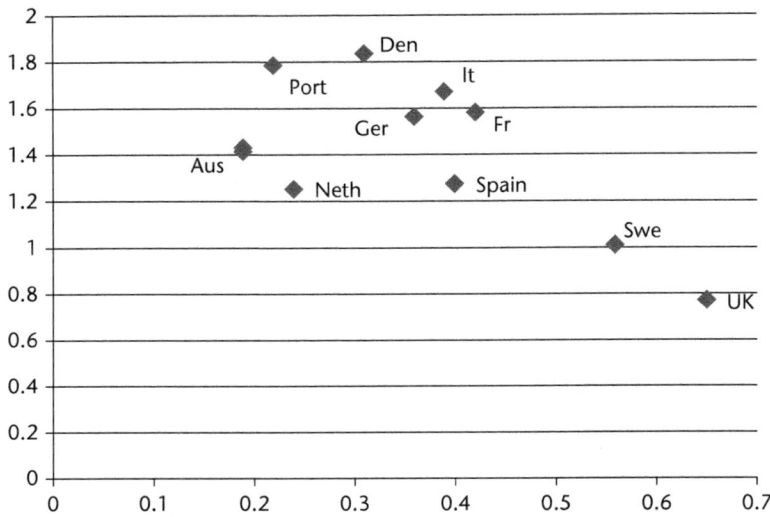

Figure 2.6. Correlation between barriers to competition at retail index and contribution of world wheat prices to retail bread prices

proxy measure that indicates broader determinants of competition. For example, the nature of planning regulations may influence the ease with which competitors (for example, food discounters) can enter the market and where they can locate. While there may be issues with this measure, it is at least available for a number of EU Member States, is measured on a consistent basis, and varies considerably across the EU.

From Figure 2.6, it is clear that the more difficult it is to enter the retail market, the less world wheat prices will influence retail prices (correlation = −0.64). This lack of influence reflects both the extent of pass-through and the magnitude of world price shocks that have had an impact on each country. This link between proxy measures for competition and the extent of price adjustment at the retail level is robust to other proxies for competition at the retail level, including barriers to entry and price controls. As a result, it seems fair to say that differences in the features of the food sector across the EU appear to be important in accounting for (or at least correlated with) the differences in food inflation in EU Member States.

In Figure 2.7, we correlate the same measure of the role of world wheat prices with the market share of hard discounters in the food sector in each of the eleven Member States. Recent reports from the EU Commission (Buckovite et al., 2009) and the European Central Bank (2011) have pointed to the role of food discounters as a characteristic of the changing structure of retail food markets in several EU Member States. In some cases, hard discounters have a strong presence (e.g., the market share of hard discounters in Germany

Food Inflation in the EU

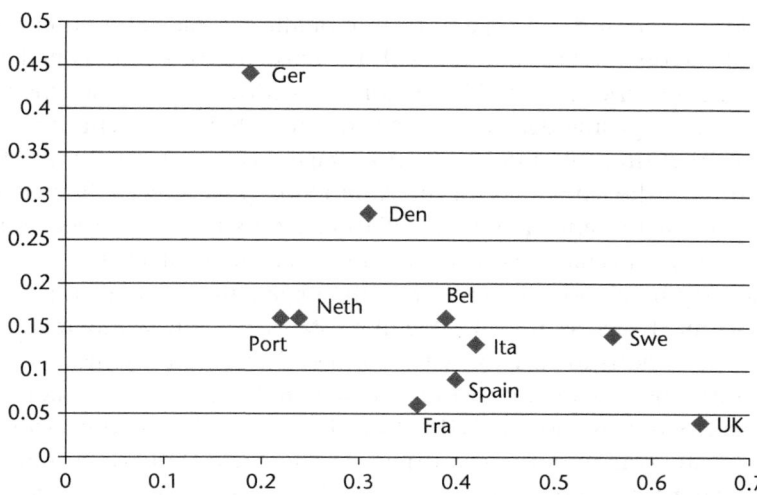

Figure 2.7. Correlation between market share of discounters and contribution of world wheat prices to retail bread prices

is 44 per cent), while in others, discounters have not experienced a significant penetration of retail food markets; in the UK and Spain, the market share is below 10 per cent. The conjecture is that the greater the presence of discounters, the more competitive the market is likely to be and hence the greater the influence of world wheat prices on the behaviour of retail bread prices.

Figure 2.7 does *not* lend support to this conjecture. While the correlation coefficient is at an acceptable level (−0.64), it does not have the sign we would anticipate for this conjecture to hold. If the presence of discounters does have an impact on how the food sector functions, it is not being reflected in the experience of how price dynamics linking retail bread and world wheat prices have behaved over time. However, the lack of (correct) correlation here is more likely to be related to a more fundamental point: given the lack of research and access to data on the food sector across EU Member States, we do not fully understand the extent to which the differences across EU Member States in the way the food chain functions matter for explaining the process of price transmission and how this is reflected in price dynamics at the retail level.

2.6 Summary and Conclusions

In this chapter, we have focused on the experience of retail food price inflation across EU Member States. In the face of commodity price shocks originating on world markets and the apparent commonality of trade and other policies that apply across the EU, together with the existence of an integrated 'single market', the experience of food inflation has varied to a considerable and to some extent surprising degree. We have documented the extent of this variation, highlighting the differences in average levels and variability of food inflation, how the experience of the commodity price shocks differed across EU Member States, and how food inflation differs from non-food inflation both in terms of average levels and variability at the retail level.

There may be obvious reasons why the experience of food inflation has varied so much across the EU. Member States rely on world markets for commodity imports to varying degrees and exchange rate regimes may offset commodity price changes on world markets. But the concern of academics, policymakers and civil society is that how the food chain functions within EU Member States is an important contributor to how retail prices adjust in face of these apparent common world price shocks. More specifically, against the background of growing concentration in the food sector and the increasingly dominant role played by food retailers, it is market power in the food sector that plays an important role in influencing the adjustment of food prices.

To this end, we also outlined the literature on price transmission with particular emphasis on vertical price transmission, which is the key focus when addressing food markets and how competition throughout the food chain may influence this process. This review supports the conjecture that competition in the food sector can influence the outcome, and given that we can observe notable differences in the structure of the food sector across the EU (and by a reasonable extension that the intensity of competition may also differ), it is not unreasonable to refer to differences in the food sector as at least part of the explanation for the differences in the experience of food inflation across EU Member States. Establishing that link, however, is more challenging given that (the lack of) competition in the food sector is multi-dimensional and difficult to establish. Although recent research has made progress on these issues, particularly in terms of progressing how price transmission should be addressed in the context of multi-product retailers, our understanding of where the exercise of market power occurs within the food supply chain (even to the benefit to consumers) and how the experience of food inflation is affected by characteristics of the food chain is still incomplete. In order to pursue the research agenda that remains, we need: (i) to access better data covering more EU Member States to understand the price transmission process, particularly at intermediate stages of the food chain; (ii) to accommodate and explicitly identify how aspects of competition in the food sector can have an impact on price transmission; (iii) to assess why the differences exist in the structure of the food sector across the EU and what are the most appropriate policies that can promote transparency in how the food chain functions.

References

Anderson, S. P., A. de Palma, and B. Krieder (2001) 'Tax incidence in differentiated product oligopoly', *Journal of Public Economics*, 81: 173–92.

Barnett, P. G., T. E. Keeler, and T.-W. Hu (1995) 'Oligopoly structure and the incidence of cigarette excise taxes', *Journal of Public Economics*, 57: 457–70.

Bettendorf, L. and F. Verboven (2000) 'Incomplete transmission of coffee bean prices: evidence from the Netherlands', *European Review of Agricultural Economics*, 27: 1–16.

Bhuyan, S. and R. A. Lopez (1997) 'Oligopoly power in the food and tobacco industries', *American Journal of Agricultural Economics*, 79: 1035–43.

Blanchard, O. and J. Gali (2007) 'The macroeconomic effects of oil shocks: why are the 2000s different from the 1970s?', NBER Working Paper No. 13368. Cambridge, MA.

Bonnet, C. and P. Dubois (2010) 'Inference on vertical contracts between manufacturers and retailers allowing for non-linear pricing and resale price maintenance', *RAND Journal of Economics*, 41: 139–64.

Bonnet, C. and V. Réquillart (2012) 'Sugar policy reform, tax policy and price transmission in the soft drink industry', TRANSFOP Working Paper No. 4 (available at <http://www.transfop.eu> accessed 14 July 2015).

Bukeviciute, L., A. Dierx, and F. Ilkovitz (2009) 'The functioning of the food supply chain and its effect on food prices in the European Union', *European Economy*, Occasional Papers No. 47. European Commission.

Cecchetti, S. and R. Moessner (2008) 'Commodity prices and inflation dynamics', *BIS Quarterly Review*, 55–66.

Delipalla, S. and O. O'Donnell (2001) 'Estimating tax incidence, market power and market conduct: the European cigarette industry', *International Journal of Industrial Organisation*, 19: 885–908.

Dixit, A. (1986) 'Comparative statics for oligopoly', *International Economic Review*, 27: 1–16.

Dornbusch, R. (1987) 'Exchange rates and prices', *American Economic Review*, 77, 93–106.

Dung, T. H. (1993) 'Optimal taxation and heterogeneous oligopoly', *Canadian Journal of Economics*, 26: 933–47.

European Central Bank (2011) 'Structural features of distributive trades and their impact on prices in the Euro Area', Structural Issues Report, September 2011. Frankfurt: European Central Bank.

EU Commission (2009) '*A Better Functioning Food Supply Chain in Europe*', Communication from the Commission to the European Parliament, the Council and the European Economic and Social Committee and the Committee of the Regions. Brussels: COM (")() 591 Final.

European Competition Network (2012) 'ECN activities in the food sector: report on competition law enforcement and market monitoring activities by European competition authorities in the food sector'. Brussels: ECN.

Feenstra, R. C. (1989) 'Symmetric pass-through of tariffs and exchange rates under imperfect competition', *Journal of International Economics*, 27: 25–45.

Gardner, B. L. (1975) 'The farm-retail spread in a competitive food industry', *American Journal of Agricultural Economics*, 57: 399–409.

Goldberg, P. and R. Hellerstein (2008) 'A structural approach to explaining incomplete exchange rate pass-through and pricing-to-market', *American Economic Review*, 98: 423–9.

Hamilton, S. F. (2009) 'Excise taxes with multi-product transactions', *American Economic Review*, 99: 458–71.

Hamilton, S. F. and T. J. Richards (2011) 'Variety and cost pass-through among supermarket retailers', Paper presented at the EAAE 2011 Congress, Zurich.

IMF (2011) *Global Economic Prospects Report, 2011*. Washington DC: IMF.

Kim, D. and R. W. Cotterill (2008) 'Cost pass-through in differentiated products: the case of US processed cheese', *Journal of Industrial Economics*, 55: 32–48.

Klenow, P. J. and L. J. Willis (2006) 'Real rigidities and nominal price changes', Working Paper No. RWP06-03, Federal Reserve Bank of Kansas.

Lloyd, T. A., S. McCorriston, C. W. Morgan, and E. Zvogu (2014) 'Dynamics of food price inflation across the EU'. Mimeo. Universities of Exeter and Nottingham.

McCorriston, S. (2002) 'Why should imperfect competition matter to agricultural economists?', *European Review of Agricultural Economics*, 29: 349–71.

McCorriston, S., C. W. Morgan, and A. J. Rayner (1998) 'Processing technology, market structure and price transmission', *Journal of Agricultural Economics*, 49: 185–201.

McCorriston, S., C. W. Morgan, and A. J. Rayner (2001) 'Price transmission: the interaction between market power and returns to scale', *European Review of Agricultural Economics*, 28: 143–59.

McCorriston, S. and I. M. Sheldon (1996) 'Trade policy reform in vertically-related markets', *Oxford Economic Papers*, 48: 664–72.

Meyer, J. and S. Von Cramon-Taubadel (2004) 'Asymmetric price transmission: a survey', *Journal of Agricultural Economics*, 55: 581–611.

Millàn, J. A. (1999) 'Market power in the Spanish food, drink and tobacco industries', *European Review of Agricultural Economics*, 26: 229–43.

Morrison-Paul, C. J. (2001) 'Market and cost structure in the beef packing industry: a plant level analysis', *American Journal of Agricultural Economics*, 83: 64–76.

Nakamura, E. and D. Zerom (2010) 'Accounting for incomplete pass-through', *Review of Economic Studies*, 77: 1192–230.

Peltzman, S. (2000) 'Prices rise faster than they fall'. *Journal of Political Economy*, 108: 466–502.

Richards, T. J., W. Allender, and S. F. Hamilton (2012) 'Commodity price inflation, retail pass-through and market power', *International Journal of Industrial Organization*, 30(1): 50–7.

Rogoff, K. S. (2009) 'Disinflation: an unsung benefit of globalisation', *Finance and Development*, 40: 54–5.

Tappata, M. (2009) 'Rockets and feathers: understanding asymmetric pricing', *RAND Journal of Economics*, 40: 673–87.

Vavra, P. and B. K. Goodwin (2005) 'Analysis of price transmission along the food chain', OECD Food, Agriculture and Fisheries Working Papers No. 3, OECD.

Walsh, J. P. (2011) 'Reconsidering the role of food prices in inflation', IMF Working Paper WP/1/71. Washington DC: IMF.

Weldegebriel, H. T. (2004) 'Imperfect price transmission: is market power really to blame?', *Journal of Agricultural Economics*, 55: 101–14.

3

Overview of Price Transmission and Reasons for Different Adjustment Patterns across EU Member States

Islam Hassouneh, Carsten Holst, Teresa Serra, Stephan von Cramon-Taubadel, and José M. Gil

3.1 Introduction

The analysis of price transmission along the food marketing chain has drawn special attention in the economics literature, as it is considered to be important to understand the overall operation of the market (Goodwin and Holt, 1999). Under a perfectly competitive market structure, economic theory shows that a price change in one market will originate similar price changes in related markets. Several empirical studies, however, have found that price dynamics differ from this theoretical competitive behaviour (von Cramon-Taubadel, 1998; Serra and Goodwin, 2003; Ben Kaabia and Gil, 2007; Serra et al., 2011; Hassouneh et al., 2012). Though a large number of studies on vertical price transmission have been conducted to characterize how prices are transmitted along the food marketing chain, little light has been shed on the factors driving different price transmission mechanisms. The objective of this chapter is to analyse price transmission processes along a wide range of different food marketing chains within the European Union (EU) and attempt to explain the differences and/or similarities that exist between them.

In characterizing price behaviour, researchers agree upon the interest in assessing the presence of asymmetries in price transmission. In the context of vertical price transmission, asymmetry usually refers to different price responses to positive and negative shocks. From a policy perspective this is an important issue, as the burden of adjustment on producers and consumers may depend on whether prices are rising or falling. A number of theoretical explanations have linked price transmission with market power (McCorriston et al., 2001; London Economics, 2004; Lloyd et al., 2006), while the specific issue for explaining asymmetric price transmission has been related to adjustment costs (Reagan and Weitzman, 1982; Azzam, 1999). Literature review articles on the causes and estimation of asymmetric price transmission are provided by Meyer and von Cramon-Taubadel (2004) and Frey and Manera (2007). Most empirical studies that explain vertical price transmission have focused on only one market. Peltzman (2000), in contrast, attempts to identify similarities in the speed with which prices respond to shocks across different US markets. He applies pre-cointegration methods to test for the existence of asymmetries in price data in a cross-market study including seventy-seven consumer goods and 165 producer goods, and finds that markets tend to respond faster to price increases than to price decreases. To explain asymmetric price behaviour, Peltzman (2000) utilizes two proxies for market power: the number of competitors and market concentration indicators. He finds conflicting results regarding market power: while a reduced number of firms is found to lead to asymmetry, higher concentration indices are found to reduce asymmetries. It is important to note, however, that the pre-cointegration techniques used by Peltzman (2000) do not take into consideration the time series properties of the data studied, which can lead to biased results. The problem of spurious regression can be avoided by applying time series procedures that are suitable for dealing with non-stationary data.

Alternative approaches to explain the causes of asymmetric price transmission have been applied: Bakucs et al. (2014) and Amikuzuno and Ogundari (2012) employ a meta-analysis. This analysis requires a well-executed systematic review and focuses on contrasting and combining the findings from independent studies to identify common patterns between these study results. Meta-analyses, however, have been widely criticized for ignoring important differences across the studies considered, such as the different methodologies, model specification, location of studies, and frequency and period of data used.

A sound comparison across different markets and countries requires the use of common methods on comparable data and period of time. This chapter aims at studying how prices are transmitted along the marketing chain of nine different food commodities and ten different EU countries. We use monthly data observed from 2000 to 2011. In contrast to Peltzman (2000), the methods used take into consideration the time series properties of data. More specifically, they allow price series to be non-stationary and to be related through a long-run equilibrium parity (i.e., to be cointegrated). A two-stage method is applied. First, a linear Vector Error Correction Model (VECM) is fitted to each market to derive the speed with which prices adjust to deviations from the long-run equilibrium relationship. In the second stage, differences in the speed of adjustment across different agricultural commodities and EU member states are assessed through the use of Tobit models. To our knowledge, this is the first research that attempts to explain differences in price behaviour across different EU markets using a fully uniform approach.

The rest of the chapter is organized as follows. A brief overview of the agro-food sector in the EU is presented in Section 3.2. Section 3.3 presents the methodological approach. The empirical results are presented in Section 3.4. Finally, the study ends with the concluding remarks section.

3.2 Agro-Food Sector in the EU

The food industry is one of the largest and most important sectors within the EU's economy. It is the second largest industry after metal, provides employment to 4.8 million people, and has an annual turnover of €917 billion. Europe's food industry is characterized by the presence of a relatively large number of small and medium-sized enterprises that operate next to a few large multinational companies (European Commission, 2013a).[1]

In 2011, the share of intra-community trade of food products was 75.1 per cent for exports and 74.3 per cent for imports.[2] The EU is considered the world's largest trader of food products. Total EU food trade with the rest of the world (the sum of extra-EU exports and imports) was €180,307 million in 2011, representing an increase of €23,146 million with respect to 2010. EU food exports reached record highs, with €76,439 million in 2010 (representing an increase of 21.8 per cent relative to 2009) and €88,982 million in 2011. EU food imports were on the order of €80,722 million in 2010 (an increase of 9.4 per cent with respect to 2009) and €91,325 million in 2011.

The EU meat industry represents nearly a quarter of the total food industry. The main meat product in the EU is pig meat (with an output of 22,387 million tons in 2011), ahead of poultry (11,201 million tons) and bovine (7,839 million tons). Germany and Spain are the main EU pig-producing countries. France and Germany are the leading cattle meat producers, while France, the UK, Germany, Poland, and Spain are the main poultry producers. In 2011, the EU produced 290,279 million tons of cereals (including rice), with wheat representing almost half of this quantity. The dairy sector is also of great importance to the EU: its cow milk production in 2011 was 138,921 million tons, of which 39 per cent came from Germany and France. The EU

---

[1] EU refers to EU-27.
[2] Unless otherwise indicated, the information presented in this section was obtained from Eurostat (2013).

is a leading exporter of dairy products, most notably cheese (European Commission, 2013b).

Since the mid-2000s, several events have motivated important changes in food prices, not only within the EU but also in international markets. The emergence of the global biofuel market, different weather effects involving substantial changes in global production, recent global food crises, and increased speculation in agricultural commodity futures markets, among other causes, have led to unprecedented changes in food price variability. An analysis aiming at assessing how these price changes are transmitted along the food marketing chain is thus timely.

3.3 Methodology

The time series econometrics literature has proposed a wide array of models to characterize price behaviour that differ in terms of sophistication and that range from simple linear VECM to very refined non-linear approaches. Convergence issues arise in the estimation process as model complexity increases. Since our priority is to compare across markets that are very different, we choose to select rather simple methodological approaches that can be fitted to most data. To achieve our objective, two different models are applied. First, we estimate a linear VECM that characterizes, in a single model, the short-run and long-run dynamics of price data. In being non-structural empirical models, VECMs restrict the type of questions that can be responded to through the analysis. As will be discussed in the results section, the linear VECM specification was chosen after ensuring that asymmetries in our price data are not relevant. Identifying the causes of the differences in the coefficients showing the adjustment of prices to deviations from the long-run parity becomes the research objective in the second stage of the analysis. A Tobit model is employed for this process. In Sections 3.3.1 and 3.3.2 we present details on the methods utilized in this study.

3.3.1 *Vector Error Correction Model*

Ignoring the properties of time series data may have important implications for conducting a sound statistical analysis of price dynamics (Myers, 1994). Two of these properties are especially relevant to our study. First, individual commodity price series generally contain stochastic trends and are non-stationary. Second, commodity prices may tend to move together over time (i.e., they may be cointegrated). In other words, though individual time series may be non-stationary, price series of interrelated markets are likely to contain the same stochastic trends. If this is the case, these stochastic trends will 'cancel' each other and a linear combination of the prices will be stationary. To build an adequate framework for price analysis, standard unit root and cointegration tests are conducted to determine whether price series are stationary and whether they are cointegrated, respectively. Since the end of the 1970s, important advances have been made in developing statistical tests for the presence of unit roots in time series. The pioneers in this field were Dickey and Fuller (1979). The augmented Dickey–Fuller (ADF) test (Dickey and Fuller, 1979) and the KPSS test (Kwiatkowski et al., 1992) are applied to each price series in order to determine whether it has a unit root (I(1)). These tests are well known and the details are therefore not discussed here.[3]

Assume that equation (1) represents the cointegration relationship between two commodity prices P_t^p and P_t^c:

$$\alpha + P_t^p - \beta P_t^c = \nu_t \tag{1}$$

where P_t^p and P_t^c are producer and consumer prices at time t, respectively, and the residual ν_t represents the deviation from the equilibrium relationship, which is often referred to as the 'Error Correction Term' (ECT). If the series are found to be cointegrated, following Engle and Granger (1987), a bivariate VECM can be expressed as follows:

$$\Delta P_t^p = \alpha_1 + \alpha^p(\alpha + P_{t-1}^p - \beta P_{t-1}^c) + \sum_{i=1}^{n}\alpha_{11}(i)\Delta P_{t-i}^p + \sum_{i=1}^{n}\alpha_{12}(i)\Delta P_{t-i}^c + \epsilon_{P_t^p}$$

$$\Delta P_t^c = \alpha_2 + \alpha^c(\alpha + P_{t-1}^p - \beta P_{t-1}^c) + \sum_{i=1}^{n}\alpha_{21}(i)\Delta P_{t-i}^p + \sum_{i=1}^{n}\alpha_{22}(i)\Delta P_{t-i}^c + \epsilon_{P_t^c} \tag{2}$$

where α and β are the parameters of the cointegration vector, ϵ_{p^p} and ϵ_{p^c} are white noise disturbances that may be correlated with each other, $\alpha_1, \alpha_2, \alpha_{11}(i), \alpha_{12}(i), \alpha_{21}(i)$ and $\alpha_{22}(i)$ are short-run dynamics parameters, and α^p and α^c are parameters that measure the rate at which prices adjust to disequilibria from the long-run cointegrating relationship. The VECM equation indicates that a change in producer (consumer) prices depend on the change in the consumer (producer) prices and also on the error correction term $(\alpha + P_{t-1}^p - \beta p_{t-1}^c)$.

Prior to fitting a linear VECM, we test for the presence of asymmetries in vertical price transmission using the model by Granger and Lee (1989). The asymmetric VECM (AVECM) can be written as follows:

---

[3] Useful surveys on unit root testing can be found in Stock (1994), Maddala and Kim (1998), Phillips and Xiao (1998), and Zivot and Wang (2005).

$$\Delta P_t^p = \alpha_1 + \alpha^{p+} v_{t-1}^+ + \alpha^{p-} v_{t-1}^- + \sum_{i=1}^{n} \alpha_{11}(i) \Delta P_{t-i}^p + \sum_{i=1}^{n} \alpha_{12}(i) \Delta P_{t-i}^c + \epsilon_{p_t^p}$$

$$\Delta P_t^c = \alpha_2 + \alpha^{c+} v_{t-1}^+ + \alpha^{c-} v_{t-1}^- + \sum_{i=1}^{n} \alpha_{21}(i) \Delta P_{t-i}^p + \sum_{i=1}^{n} \alpha_{22}(i) \Delta P_{t-i}^c + \epsilon_{p_t^c} \quad (3)$$

where $v_t^+ = v_t$, $v_t = \alpha + P_{t-1}^p - \beta p_{t-1}^c$, for all $v_t > 0$ and 0 otherwise and $v_t^- = v_t$ for all $v_t < 0$ and 0 otherwise. Since $v_t^+ + v_t^- = v_t$, the standard symmetric VECM is nested in the AVECM and an F-test can be used to test the null hypothesis of symmetry ($\alpha^{p+} = \alpha^{p-}$). Since ample evidence against asymmetric price behaviour was found we choose to use a linear VECM.

3.3.2 Tobit Model

To understand the differences in price adjustments across different countries and commodities, two Tobit models are estimated, one with the producer price adjustment to long-run disequilibrium as a dependent variable and another showing the same adjustment but for the consumer price. In a cointegrated market, where producer and consumer prices respond to deviations from the long-run parity, $\alpha^p < 0$ and $\alpha^c > 0$ (see equation (2)). For ease of interpretation of Tobit model research results, $\alpha^p < 0$ is taken in absolute values. This leads both the producer and the consumer models to be left censored.

The variables that have been selected as explanatory variables in the Tobit model are necessarily restricted by data availability, since we need these variables to be available for different EU countries and for different agro-food sectors. The specification of the two Tobit models is as follows:

$$\alpha^p = \beta^P + \beta_1^P PSR_i + \beta_2^P ESR_i + \beta_3^P PS_i + \beta_4^P AUS_i + \beta_5^P FRA_i + \beta_6^P GER_i$$
$$+ \beta_7^P UK_i + \beta_8^P CHI_i + \beta_9^P EGG_i + \beta_{10}^P POR_i + \beta_{11}^P MIL_i + \beta_{12}^P BUT_i$$
$$+ \beta_{13}^P CHE_i + \beta_{14}^P BRE_i + \beta_{15}^P FLO_i + u_i^p$$

$$\alpha^c = \beta^C + \beta_1^C PSR_i + \beta_2^C ESR_i + \beta_3^C PS_i + \beta_4^C AUS_i + \beta_5^C FRA_i + \beta_6^C GER_i$$
$$+ \beta_7^C UK_i + \beta_8^C CHI_i + \beta_9^C EGG_i + \beta_{10}^C POR_i + \beta_{11}^C MIL_i + \beta_{12}^C BUT_i$$
$$+ \beta_{13}^C CHE_i + \beta_{14}^C BRE_i + \beta_{15}^C FLO_i + u_i^c \quad (4)$$

where α^p and α^c are the producer and consumer price speed of adjustments to long-run disequilibrium, respectively. PSR_i is the production specialization ratio and represents the ratio of a country's production of a specific product to total meat or non-meat production of the same country, depending on the type of commodity considered (i.e., the beef specialization ratio is measured as the amount of bovine meat divided by the gross meat production). ESR_i is the export specialization ratio defined as the share of total country exports of a specific product divided by the total meat or non-meat exports of the same country. PS_i is the proportion of country production of a specific product over the total EU production of the same product. AUS_i, FRA_i, GER_i, and UK_i are country dummy variables for Austria, France, Germany, and the UK, respectively. They account for country differences in price behaviour. While different country dummies were considered, we only kept those dummies leading to statistically significant coefficients. The rest of the variables considered are commodity dummies that control for any significant differences between price behaviour in different food marketing chains. Specifically, we consider poultry, eggs, pork, liquid milk, butter, cheese, bread, and flour marketing chains (POU_i, EGG_i, POR_i, MIL_i, BUT_i, CHE_i, BRE_i, and FLO_i, respectively). Price behaviour in the beef market is chosen as the benchmark against which to compare the rest of the products. Variable u represents the stochastic disturbance term and i is the ith observation.

3.4 Empirical Results

This chapter aims to conduct an analysis to characterize price transmission processes along a wide range of different food marketing chains within the EU, and to explain the differences and/or similarities that exist between them. To achieve this objective, different commodity price data across different European countries were collected. The range of products and countries considered is limited by data availability. These data were collected within the framework of the TRANSFOP project (<http://www.transfop.eu/>) and consist of monthly data over the period from January 2000 to December 2011, obtained from different national statistical sources.[4] Specifically, we analyse price transmission within the beef, eggs, poultry, pork, milk, butter, cheese, bread, and flour industries in ten EU Member States (Austria, Belgium, France, Germany, Hungary, Italy, Slovakia, Slovenia, Spain, and the UK). Graphs of poultry price series for the selected ten EU countries are presented in Figure 3.1.[5]

---

[4] More details regarding the data used are available from the authors upon request.
[5] Poultry prices are presented since they are available for all countries studied.

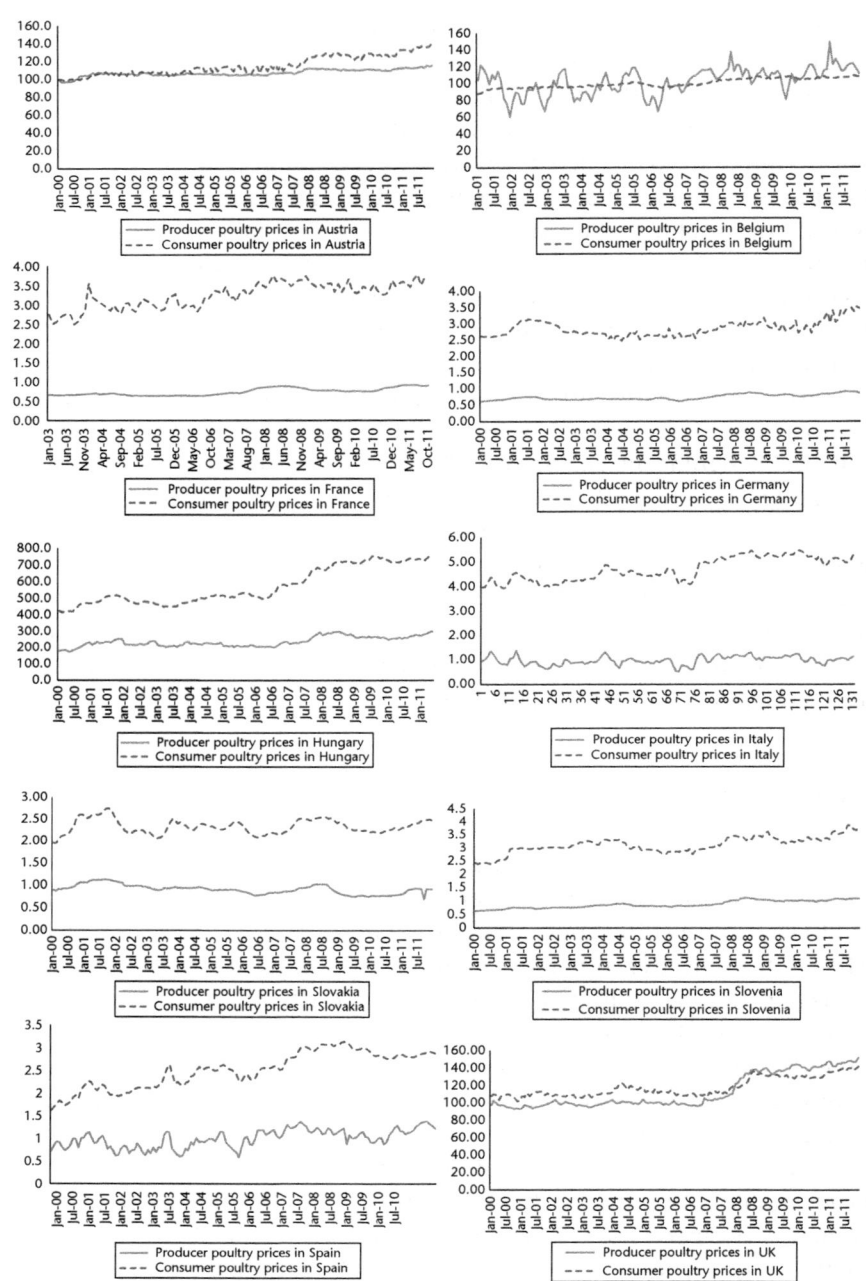

Figure 3.1. Poultry price series for ten EU countries

Price Transmission and Adjustment Patterns in the EU

The empirical analysis is based on a logarithmic transformation of time series data. Standard unit root and Engle and Granger (1987) cointegration tests confirm the presence of a unit root and provide evidence of a long-run equilibrium relationship between most producer and consumer prices.[6] As noted above, error correction models capture both the short- and long-run dynamics of non-stationary and cointegrated data. Prior to fitting a linear VECM to price data, the presence of asymmetries is tested for by estimating an asymmetric VECM (AVECM) for each market considered. Unlike Peltzman (2000), who found strong support for asymmetries in US food markets, and consistent with the economic theory, our results indicate that 66 per cent of the price transmission processes studied are symmetric.[7] As a result, a linear VECM is chosen as the most adequate model specification.

The estimated VECMs are bi-dimensional and have producer and consumer prices as endogenous variables. Alpha parameters in our VECM provide information on the speed of adjustment of prices to disequilibria from the long-run parity that exists between prices at different levels of the food marketing chain. Results of the VECMs are presented in Table 3.1.[8] Parameter estimates suggest that, for the majority of markets studied, producer prices tend to adjust to deviations from the equilibrium prices at a faster speed than consumer prices. Specifically, producer and consumer average price response to long-run disequilibrium is 11 per cent and 3 per cent, respectively. The overall speed of adjustment (the sum of producer and consumer response) provides evidence that eggs, poultry, and pork respond more quickly to deviations from the long-run equilibrium compared to beef prices and dairy and wheat markets. Beef market prices in France and Slovakia show no tendency to converge towards the long-run equilibrium parity.

Our specific objective is to explain the differences in alphas across different countries and commodities. Tobit model results are presented in Table 3.2 and indicate that an increase in the degree of country specialization in producing a certain food product reduces the speed of producer price adjustment to market price disequilibria (48 per cent). Increased specialization can furnish a basis for producer market power, which in turn can make producer stickier prices or slowly responsive to market shocks. Results show that the export

---

[6] Results of standard unit root and cointegration tests are not presented here, but are available from the authors upon request.

[7] While commodity prices at producer and consumer level in Austria do not present any asymmetric adjustment, most commodities in Belgium adjust in a non-linear fashion. Other markets lie between these two countries. Findings also indicate that perishable products are more likely to present asymmetric adjustment than other commodities. Moreover, parameter estimates show that for perishable products (eggs, milk, and cheese) consumers are more likely to adjust to deviations from the long-run equilibrium relationship. Results of AVECM are available from the authors upon request.

[8] The optimum number of lags in specifying the VECM is selected by using the SBC and AIC criterion. To preserve space, only parameters showing the long-run adjustments are presented.

Table 3.1. Estimation of the Vector Error Correction Model: Adjustment of producer and consumer prices to long-run disequilibrium

Commodity/Country	Independent variable	Austria P.E.	Austria C.E.	Belgium P.E.	Belgium C.E.	France P.E.	France C.E.	Germany P.E.	Germany C.E.	Hungary P.E.	Hungary C.E.	Italy P.E.	Italy C.E.	Slovakia P.E.	Slovakia C.E.	Slovenia P.E.	Slovenia C.E.	Spain P.E.	Spain C.E.	UK P.E.	UK C.E.
Beef	ECT_{t-1}	−0.144** (0.030)	0.021* (0.011)	−0.084** (0.030)	0.012* (0.006)	−0.080 (0.049)	0.089 (0.054)	−0.130** (0.036)	0.034** (0.015)			−0.131** (0.038)	0.055** (0.015)	−0.041 (0.025)	0.014 (0.016)			−0.118** (0.041)	0.002 (0.012)	0.010 (0.040)	0.103** (0.018)
Eggs	ECT_{t-1}			−0.293** (0.066)	0.013** (0.006)					−0.437** (0.097)	−0.042 (0.035)									−0.124** (0.039)	0.016 (0.017)
Milk-Butter	ECT_{t-1}			−0.092** (0.022)	0.007 (0.007)			−0.083** (0.023)	0.026 (0.060)	−0.062** (0.022)	0.046** (0.016)	−0.089** (0.031)	0.011 (0.014)								
Milk-Cheese	ECT_{t-1}			−0.072** (0.025)	0.013** (0.003)			−0.040** (0.017)	0.068** (0.021)	−0.103** (0.044)	0.071** (0.042)	−0.044 (0.029)	0.039** (0.011)			−0.066** (0.029)	0.031 (0.042)				
Milk-Milk	ECT_{t-1}	−0.063** (0.023)	−0.009 (0.011)	−0.082** (0.020)	0.005 (0.007)	−0.246** (0.055)	0.023** (0.009)	−0.057** (0.022)	0.074** (0.030)	−0.074** (0.024)	0.026 (0.018)	−0.048 (0.032)	0.024** (0.014)	−0.035** (0.016)	0.035** (0.011)	−0.065** (0.031)	0.056** (0.028)	−0.057** (0.026)	0.031** (0.009)	−0.157** (0.034)	0.008 (0.017)
Pork	ECT_{t-1}			−0.258** (0.055)	0.015** (0.007)	−0.201** (0.063)	0.046 (0.034)	−0.064 (0.058)	0.091** (0.014)	−0.165** (0.062)	−0.006 (0.037)	−0.277** (0.053)	0.007 (0.014)	−0.317** (0.088)	0.005 (0.034)			−0.236** (0.073)	−0.005 (0.016)	−0.136** (0.044)	0.026 (0.038)
Poultry	ECT_{t-1}	−0.106** (0.041)	0.381** (0.134)	−0.315** (0.066)	0.025** (0.006)	−0.020 (0.018)	0.194** (0.064)	−0.026 (0.019)	0.101** (0.045)	−0.140** (0.046)	−0.033 (0.022)	−0.469** (0.080)	−0.020 (0.015)			0.011 (0.008)	0.031** (0.010)			0.000 (0.028)	0.139** (0.031)
Wheat-Bread	ECT_{t-1}			−0.042 (0.026)	0.009** (0.004)			−0.054** (0.020)	0.003 (0.003)	−0.089** (0.033)	0.013** (0.005)			−0.084** (0.027)	0.011** (0.004)	−0.123** (0.051)	0.008 (0.012)				
Wheat-Flour	ECT_{t-1}	−0.062* (0.037)	0.030** (0.007)	0.002 (0.028)	0.031** (0.007)					−0.121** (0.048)	0.045** (0.014)					−0.169** (0.058)	0.050** (0.019)	0.009 (0.020)	0.019** (0.005)		

Notes: Numbers in parentheses are standard errors.
** and * denote statistical significance at the 5 and 10 per cent significance level, respectively.
P. E. = Producer equation, C. E. = Consumer equation.

Table 3.2. Tobit results: Parameter estimates

Independent variable	Producer equation	Consumer equation
PSR	−0.482** (0.141)	0.119 (0.102)
ESR	−0.912** (0.232)	0.357** (0.172)
PS	1.366** (0.357)	−0.495* (0.288)
AUS	−0.040 (0.048)	0.102** (0.039)
FRA	−0.172** (0.061)	0.070 (0.049)
GER	−0.193** (0.063)	0.085* (0.049)
UK	−0.215** (0.058)	0.047 (0.044)
POU	0.085* (0.046)	0.070* (0.035)
EGG	0.202** (0.061)	−0.016 (0.055)
POR	0.308** (0.065)	−0.075 (0.054)
MIL	0.062 (0.045)	−0.007 (0.035)
BUT	−0.126* (0.064)	0.004 (0.056)
CHE	−0.095* (0.055)	0.055 (0.045)
BRE	0.081 (0.061)	−0.034 (0.050)
FLO	0.040 (0.057)	0.008 (0.044)
β	0.190** (0.047)	−0.046 (0.038)
Number of observations	56	56
P-value	0.000	0.045

Notes: Numbers in parentheses are standard errors.
** and * denote statistical significance at the 5 and 10 per cent significance level, respectively.
PSR = production specialization ratio, ESR= export specialization ratio, PS = production share within the EU, AUS = Austria, FRA = France, GER = Germany, UK = UK, POU = poultry, EGG = eggs, POR = pork, MIL = milk, BUT = butter, CHE = cheese, BRE = bread, FLO = flour, β = constant.

specialization ratio also reduces the need for producer prices to adjust to the long-run disequilibrium. A wider array of market outlets may reduce producers' need to quickly adjust prices to market shocks. In contrast, national retail prices are found to adjust more quickly (36 per cent) within growing export specialization. This is likely due to the fact that international competition increases the pressure on the national prices paid by consumers and forces quicker adjustments in a usually rigid market. The model provides evidence that a country's share within total EU production implies faster producer adjustment coefficients. In other words, the more relevant the production is, the higher the need to sell agricultural commodities and thus the quicker the adjustment of producer prices. On the other hand, an important supply of commodities may be beneficial to retailers, which translates into a more prolonged adjustment path.

The analysis also suggests different price adjustment in different EU countries. Specifically, results show that France, Germany, and UK producer prices adjust less in comparison to other EU countries. In contrast, Austria and Germany have higher consumer adjustment rates compared to other EU countries. Regarding commodity dummies, the producer price equation indicates that poultry, eggs, and pork adjust more quickly to deviations from the long-run equilibrium compared to beef prices (the sector taken as reference in specifying the commodity dummies). These results are consistent with VECM findings. Quicker adjustments are also registered in the poultry retail price. This may be due to the highly industrialized production processes that concentrate most of the production of these commodities, as well as the vertically integrated structures that characterize these industries and that can facilitate the process of price adjustment. The observed milder producer adjustment in the processed milk products (butter and cheese) may be due to their long life relative to fresh milk, which reduces the pressure to sell the commodity in the short term.

3.5 Concluding Remarks

Price transmission analyses using time series data are usually conducted through the use of econometric non-structural models that provide a characterization of price behaviour, but that do not provide any objective explanation of the reasons behind it. The researcher has to infer these underlying causes based on his/her knowledge of the sector studied. Explaining price behaviour becomes increasingly difficult when a wide range of markets is taken into consideration. This chapter aimed to shed light on this issue by conducting an analysis to characterize price transmission processes along a series of different food marketing chains within the EU, and attempted to explain the differences and/or similarities that exist between them.

To achieve this objective, two different techniques were considered. After testing for asymmetries in vertical price transmission and finding strong evidence of linear price behaviour for the time span and markets considered, a linear VECM was used to capture how prices are transmitted among producer and consumer markets. Tobit models were then applied to explain the differences in these adjustment patterns. The parameter estimates from the VECMs indicate that most upstream prices in the marketing chain adjust faster than consumer prices that are sticky and slowly responsive to market shocks.

Tobit results suggest that an increase in production and export specialization ratios, which may bring increased market power, reduces the speed of producer price adjustments. Conversely, national retailer prices adjust more quickly when market outlets for agricultural producers increase. Findings also suggest that an increase in the share of a country's national production within the EU's production leads to quicker adjustments in producer markets. In addition, vertically integrated supply chains are more prone to show quick adjustment to long-run disequilibria compared to less integrated ones. Highly perishable products are also seen to have more responsive prices than products with longer shelf lives.

The main shortcoming of our analysis is the restricted range of variables considered in explaining price behaviour differences. The range of explanatory variables is limited by data availability. In this regard, increased availability of homogeneous information across EU countries would allow for refining research results and would contribute to shedding more light on the reasons behind the differences in price adjustment mechanisms.

References

Amikuzuno, J. and K. Ogundari (2012) 'The contribution of agricultural economics to price transmission analysis and market policy in sub-Saharan africa: what does the literature say?', Contributed paper, 86th Annual Conference of the Agricultural Economics Society, University of Warwick, United Kingdom.

Azzam, A. M. (1999) 'Asymmetry and rigidity in farm-retail price transmission', *American Journal of Agricultural Economics*, 81: 525–33.

Bakucs, Z., J. Falkowski, and I. Ferto (2014) 'Does market structure influence price transmission in the agro-food sector? a meta-analysis perspective', *Journal of Agricultural Economics*, 65: 1–25.

Ben Kaabia, M. and J. M. Gil (2007) 'Asymmetric price transmission in the Spanish lamb sector', *European Review of Agriculture Economics*, 34: 53–80.

Dickey, D. A. and W. A. Fuller (1979) 'Distribution of the estimators for autoregressive time series with a unit root', *Journal of the American Statistical Association*, 74: 427–31.

Engle, R. F. and C. W. J. Granger (1987) 'Co-integration and error correction: representation, estimation and testing', *Econometrica*, 55: 251–76.

European Commission (2013a) 'Food industry, EU food market overview'. Accessed February 2013, available at <http://ec.europa.eu/enterprise/sectors/food/eu-market/index_en.htm>.

European Commission (2013b) 'Agriculture and rural development, milk and milk products'. Accessed February 2013, available at <http://ec.europa.eu/agriculture/milk/index_en.htm>.

Eurostat (2013) Dataset. Accessed February 2013, available at <http://epp.eurostat.ec.europa.eu/portal/page/portal/statistics/themes> accessed 1 February 2013.

Frey, G. and M. Manera (2007) 'Econometric models of asymmetric price transmission', *Journal of Economic Surveys*, 21: 349–415.

Goodwin, B. K. and M. Holt (1999) 'Price transmission and asymmetric adjustment in the U.S. beef sector', *American Journal of Agricultural Economics*, 81: 630–7.

Granger, C. W. J. and T. H. Lee (1989) 'Investigation of production, sales and inventory relationships using multicointegration and non-symmetric error correction models', *Journal of Applied Econometrics*, 4: 145–59.

Hassouneh, I., A. Radwan, T. Serra, and J. M. Gil (2012) 'Food scare crises and developing countries: the impact of avian influenza on vertical price transmission in the Egyptian poultry sector', *Food Policy*, 37: 264–74.

Kwiatkowski, D., C. B. Phillips, P. Schmidt, and Y. Shin (1992) 'Testing the null hypothesis of stationarity against the alternative of a unit root: how sure are we that economic time series have a unit root?', *Journal of Econometrics*, 54: 159–78.

London Economics (2004) 'An investigation of the determinants of farm-retail price spreads', Final Report to DEFRA by London Economics, February 2004. London.

Lloyd, T. A., S. McCorriston, C. W. Morgan, and A. J. Rayner (2006) 'Food scares, market and price transmission: the UK BSE crisis', *European Review of Agricultural Economics*, 33: 119–47.

McCorriston, S., C. W. Morgan, and A. J. Rayner (2001) 'Price transmission: the interaction between market power and returns to scale', *European Review of Agricultural Economics*, 28: 143–59.

Maddala, G. S. and I. M. Kim (1998) *Unit Roots, Cointegration and Structural Change*. Oxford: Oxford University Press.

Meyer, J. and S. von Cramon-Taubadel (2004) 'Asymmetric price transmission: a survey', *Journal of Agricultural Economics*, 55: 581–611.

Myers, R. J. (1994) 'Time series econometrics and commodity price analysis: a review', *Review of Marketing and Agricultural Economics*, 62: 167–81.

Peltzman, S. (2000) 'Prices rise faster than they fall', *Journal of Political Economy*, 108: 466–502.

Phillips, P. C. B. and Z. Xiao (1998) 'A primer on unit root testing', *Journal of Economic Surveys*, 12: 423–70.

Reagan, P. and M. Weitzman (1982) 'Asymmetries price and quality adjustments by the competitive firm', *Journal of Economic Theory*, 27: 410–20.

Serra, T. and B. K. Goodwin (2003) Price transmission and asymmetric adjustment in the Spanish dairy sector', *Applied Economics*, 35: 1889–99.

Serra, T., D. Zilberman, J. M. Gil, and B. K. Goodwin (2011) 'Nonlinearities in the US corn-ethanol-crude oil price system', *Agricultural Economics*, 42: 35–45.

Stock, J. H. (1994) 'Unit roots, structural breaks and trends', in R. F. Engle and D. L. McFadden (eds), *Handbook of Econometrics*, vol. 4: 2739–841. New York: North Holland.

von Cramon-Taubadel, S. (1998) 'Estimating asymmetric price transmission with the error correction representation: an application to the German pork market', *European Review of Agricultural Economics*, 25: 1–18.

Zivot, E. and J. Wang (2005) *Modeling Financial Time Series with S-plus*. New York: Springer.

4

Price Transmission in Food Chains: The Case of the Dairy Industry

Céline Bonnet, Tifenn Corre, and Vincent Réquillart

4.1 Introduction

On average, the share of household expenditure on food in Europe was about 13 per cent in 2011 (Eurostat data). There is, however, a large variation in expenditure shares across EU countries. For example, the share of household expenditure on food in 2011 was 10.2 per cent in Ireland, 13.7 per cent in France, and 18.5 per cent in Poland. More importantly, because food is a staple good, there is a large variation across income levels. Thus, whereas the average percentage of household expenditure on food was 16.8 per cent in 2005, it was 22.2 per cent for the first income quintile and 13.0 per cent for the fifth quintile.[1] In a context in which a significant portion of the EU population is concerned about food price inflation, this raises the question of how shocks in agricultural prices are transmitted to consumers' prices. As shown by Bukeviciute et al. (2009), following the peak of agricultural commodity prices in 2007–8, food price inflation in the EU has displayed considerable discrepancies across countries. For example, the elasticity of consumer food prices to producer food prices was about 15–20 per cent in the Eurozone on average but reached 30 per cent in Sweden. Lloyd et al. (2013) point out that the heterogeneity of price transmission in the EU is linked to the strong heterogeneity of how the food chain functions in the different Member States and for the different food chains.

---

[1] <http://epp.eurostat.ec.europa.eu/portal/page/portal/household_budget_surveys/Data/database> (accessed 12 April 2014).

Due to the Common Agricultural Policy (CAP) reforms in the 2000s, the EU prices of agricultural products now experience larger variations than in the past. These reforms dramatically changed the way in which farmers' incomes are supported in the EU. Whereas support was mostly administered through price support in the 1990s, most of the support is now provided through direct payments. Thus, in 1990, price support represented 84 per cent of the support in the EU, whereas it was only 12 per cent in 2011.[2] Moreover, because the tools to avoid large price decreases were mainly dismantled, prices can experience greater decreases than in the past. As an example, the milk price dropped by 30 per cent in 2009, whereas this had not happened in the previous fifteen years. In addition, EU prices for agricultural products are now more connected to world prices, and have been since some of the forms of protection that isolated the EU agricultural markets from the rest of the world were removed. Thus, the Producer Nominal Protection Coefficient, which is the ratio of the average price received by producers at the farm gate to the border price, was 1.48 in 1990 but had fallen to 1.03 in 2011.

A standard way to analyse price transmission along the food supply chain is to base the analysis on time series. For example, Hassouneh et al. (2013) developed a systematic analysis of price transmission in the EU. They documented the heterogeneity of price transmission among countries and commodities. They found that producer prices tend to adjust to deviations from equilibrium prices at a faster speed than consumer prices. They also found that some commodities (eggs, poultry, and pork) respond more quickly to deviations from the long-run equilibrium price compared to other commodities (beef, dairy products, wheat products). Moreover, they found that an increase in the degree of a country's specialization in producing a certain food product reduces the speed of producer price adjustment to market price disequilibria. In the case of dairy markets in Germany, Loy et al. (2015) showed that the price adjustment of private labels (PLs) was much faster than that of national brands (NBs). On the whole, these methods allow for the characterization of price transmission in different markets, but the main determinants of food price transmission remain unclear (for a recent survey, refer to Frey and Manera, 2007).

An alternative way to analyse price transmission is to use a structural model in order to better understand the behaviour of the food supply chain and to deduce the implications for price transmission (recent contributions include Kim and Cotterill, 2008; Nakamura and Zerom, 2010; Bonnet et al., 2013). This methodology is appropriate in the case of concentrated markets in which firms are able to strategically set prices. This is the case in numerous food

---

[2] OECD PSE database, <https://stats.oecd.org/Index.aspx?DataSetCode=MON20123_1> (accessed 15 April 2014).

supply chains, which typically consist of large firms with significant market power. In the food industry, the market share of the top three manufacturers is frequently greater than 50 per cent and the top five retailers now account for over 50 per cent of the grocery market in many EU countries (McCorriston, 2013). Food supply chains are therefore frequently composed of a chain of oligopolies.

In a context of perfect competition, cost pass-through is lower than or equal to 1 and depends on the elasticities of supply and demand.[3] In a context of imperfect competition, cost pass-through also depends on markup adjustments. In particular, the literature on taxation under conditions of imperfect competition has shown that the cost pass-through might be less than or greater than 1 depending on the curvature of the demand function (Stern, 1987; Delipalla and Keen, 1992; Anderson et al., 2001).

For example, Nakamura and Zerom (2010) studied the US coffee industry and reported a long-run pass-through of coffee commodity prices to retail prices of 0.92. Recently Bonnet and Réquillart (2013) found that the cost pass-through in the French soda market is 1.16, on average. Hellerstein (2008) showed that markup adjustments at the manufacturer and retailer levels play an important role in explaining the pass-through of cost changes in the US beer industry. Overall, these studies suggest that firms should strategically adjust their markups when facing a change in their input costs. Moreover, as shown by Bonnet et al. (2013), the pass-through rate for upstream cost shocks to downstream retail prices depends on the form of the contracts between manufacturers and retailers. This literature suggests that to assess price transmission along a particular food supply chain, it is necessary to consider key characteristics such as the structure of the chain, consumers' substitution patterns, and the type of contracts linking manufacturers and retailers.

In this chapter, we develop a structural econometric model that allows for the assessment of the price transmission of a cost change, taking into account horizontal and vertical interactions between manufacturers and retailers. We follow the methodology developed by Berto Villas Boas (2007) and Bonnet and Dubois (2010) and recently used by Bonnet and Réquillart (2013). This methodology mainly consists of a three-step analysis. In the first step, a demand model is designed and estimated. Then, in the second step, given the demand estimates, price–cost margins are computed for a set of contracts between food manufacturers and retailers. For each type of contract, the first-order conditions for profit maximization allow for definition of structural equations that define price–cost margins in the chain. This step also

---

[3] Following Kim and Cotterill (2008), we assume that the 'cost pass-through rate is defined as the proportion of a change in input cost that is passed through to the final price of the product'.

includes the selection of the model of vertical relationships that best fits the data. Finally, the third step consists of using the selected model to perform simulations.

We apply this methodology to two dairy markets: the fluid milk market and the dairy desserts market. We choose those industries for many reasons. First, the milk industry as a whole is one of the main food industries in Europe, comprising 14 per cent of the annual turnover of all food industries (Food Drink Europe, 2014). Second, it encompasses final products that present different characteristics in terms of concentration, PL penetration, and cost of raw milk in the production process. Third, as explained above, the farm milk price has experienced large price variations in the recent past. We show that the pass-through for NBs is greater than 1 in the fluid milk market, whereas it is less than 1 in the dairy desserts market. Therefore, depending on the market, firms will over-shift or under-shift cost changes. It should be noted that in both markets, we found that PLs fully transmit the cost change (the pass-through is very close to 1 for all PLs whatever the market considered).

This chapter is organized as follows. Section 4.2 presents the data and provides descriptive statistics about dairy markets. Section 4.3 describes the model and methods used to analyse consumers' demand and to infer the vertical relationships between manufacturers and retailers. Section 4.4 discusses the results in relation to the demand for differentiated dairy products, the role of vertical relationships in both industries, and how they determine cost pass-through. Finally, Section 4.5 concludes.

4.2 French Dairy Market and Data

Before 2006, the farm milk price in France was rather predictable (Figure 4.1). From 1990 to 2001, the average annual price was roughly constant and the monthly price followed a clear seasonal cycle, with high prices in winter and lower prices in summer. From 2001 to 2007, the seasonal pattern did not change but the average annual price followed a negative trend that is easily explained by the reform of the dairy sector, which consisted of lowering support prices and increasing quotas.[4] Since 2007, the milk price has experienced a period of higher volatility, with a peak in 2008 and a low in 2009, which is explained by the removal of the main tools

---

[4] In the framework of the Luxembourg reform (2003), the intervention price for butter falls by 25% in four steps from 2004 to 2007 and the intervention price for skimmed milk powder falls by 15% in three steps from 2004 to 2006. For an analysis of the milk dairy reform, refer to Bouamra-Mechemache et al. (2008).

Price Transmission in Food Chains: The Dairy Industry

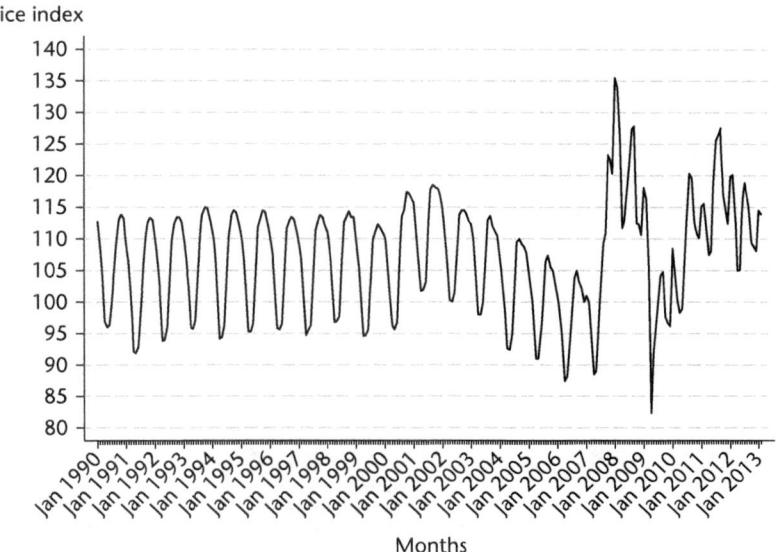

Figure 4.1. Raw milk price in France 1990–2011

that were used to support the farm milk price. It is also explained by the growing influence of world market prices on EU prices. In this study, we focus on the last period (2006–09) to analyse, in the context of higher volatility, how changes in the farm milk price are transmitted through to retail prices.

To investigate the impact of raw milk price changes on consumer prices, we study two final markets: the dairy desserts market and the fluid milk market. As explained in the introduction, we choose these markets because their structure, the importance of differentiated products, the importance of PLs, and the cost structure differ significantly. In particular, according to the 'Observatoire des prix et des marges', the raw milk cost is about one-third of the consumer price for liquid milk and one-sixth for yoghurt (a type of dairy dessert) and it is likely to be much less for more processed desserts such as cream desserts.[5]

In the case of dairy desserts (Figure 4.2), there is a link between retail and farm milk prices, but for non-fat products, the link seems to be smaller than it is for regular products. For example, for yoghurts, the correlation between the raw milk price and the retail price of yoghurts is 0.75 for standard yoghurt and it is −0.01 for non-fat yoghurts. In the case of the fluid milk market, there is a clear link between the evolution of retail price changes and the evolution of the raw milk price. The changes in the skimmed milk price are smaller than for

---

[5] <https://observatoire-prixmarges.franceagrimer.fr/resultats/Pages/ResultatsFilieres.aspx?idfiliere=6> (accessed 12 May 2014).

Céline Bonnet, Tifenn Corre, and Vincent Réquillart

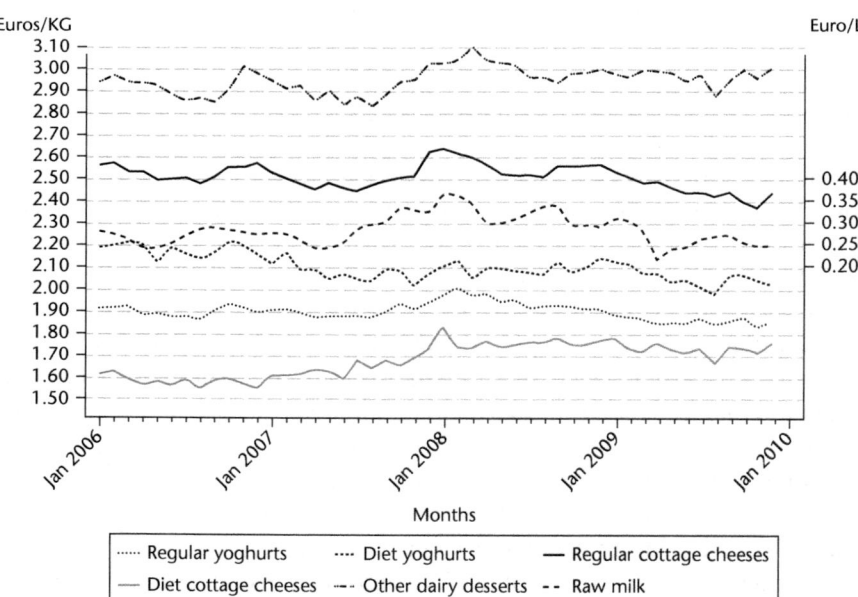

Figure 4.2. Evolution of prices 2006–09

semi-skimmed milk and whole milk, but the general evolution of prices suggests a strong link between farm and retail prices.

4.2.1 Data

To conduct the analysis of both sectors, we use consumer panel data collected by *Kantar Worldpanel*, a French representative survey of 20,000 households, over a four-year period (2006–09). This survey records information about every purchase of food products (e.g., quantity, price, brand, characteristics of goods, and the retailer from which products are purchased) for each household in the panel. The data only cover household purchases for home consumption, and thus out-of-home consumption is excluded. It should be noted, however, that home consumption dominates, as home consumption accounts for 86.5 per cent of total consumption of dairy desserts and 95 per cent of total consumption of fluid milk (AFSSA, 2009).[6]

From a consumer perspective, a product is the combination of a brand and a retailer. Thus, brand 1 bought at retailer 1 is different from brand 1 purchased at retailer 2. This distinction allows us to analyse vertical contracts between manufacturers and retailers as well as different price strategies between retailers. As shown by Steiner (1993), the balance of power between manufacturers and retailers within a chain is strongly related to the loyalty of consumers towards brands or stores. It is therefore important to characterize a purchase in terms of both the brand purchased and the retailer from which the consumer purchased the product. Retailers are grocery store chains that differ by both the size of their outlets and the services they provide to consumers. We consider the top five retailers that operate in the French retail sector (three of them are characterized by large outlets, while the other two have intermediate-sized outlets) and two aggregates. One aggregate includes the discounters, which have small to intermediate-sized outlets and provide basic services only. The other aggregate comprises the remaining retailers. All these retailers are national chains present throughout France. Consumers from the different regions in France are presented with the same assortment of products when they visit a given retailer.[7] We now provide specific information on brands for the two markets.

4.2.2 The Market for Dairy Desserts

Dairy desserts are a part of the dessert market. We consider five product categories of dairy desserts, and to deal with the substitution that occurs

---

[6] EtudeIndividuelle des Consommations Alimentaires INCA2, <http://www.anses.fr/sites/default/files/documents/PASER-Ra-INCA2.pdf> (accessed 6 June 2014).
[7] This is true for the NBs we consider, which are the main ones. It might be the case that in some small outlets, the assortment is reduced.

with other desserts, we define an 'outside option' comprised of pastries, ice creams, and fruits. The five product categories are defined as follows. First, we define three segments of dairy desserts: yoghurts (plain, flavoured, or fruit yoghurts), cottage cheeses, and other dairy desserts such as cream desserts, creamy rice puddings, mousses, or custards. In addition, for the first two segments, we distinguish low-fat products from regular products. We thus have five product categories. On the whole, the five product categories represent about 53 per cent of the entire market.[8] According to our sample, Danone, Yoplait, and Nestlé are the three main manufacturers within the market for dairy desserts. They represent 42 per cent of purchases of these products, whereas the market share of PLs is 49 per cent. The remaining 9 per cent of the market is covered by other manufacturers producing NBs. These products are gathered in an aggregate. To assess price formation at the brand level, we consider the four main brands produced by Danone (Activia, Danette, Taillefine, and Gervais), the main brand for Yoplait (Panier de Yoplait), and the main brand for Nestlé (La laitire). We also define three aggregates, each one including brands of Danone, Yoplait, and Nestlé other than those described previously. Finally, we consider five PLs, one for each of the five product categories previously described. Given this segmentation by brands and by type of products, we finally have twenty-three brands with eighteen NBs and five PLs. Because each of these products is available in each of the seven retailers, we thus consider 159 differentiated products that compete in the market for dairy desserts.[9]

The average price of dairy desserts is €2.27/kg (Table 4.1).[10] The average prices of regular yoghurts, regular cottage cheeses, and other dairy desserts are €1.94, €2.42, and €2.96/kg, respectively. The main market segment is yoghurt (58 per cent of dairy dessert purchases). Regular products dominate the market since they represent roughly 85 per cent of all dairy dessert purchases. On average, for each product category, NBs' prices are 50 per cent higher than PLs' prices. Yoghurts are the cheapest products, for both types of brands, followed by cottage cheese and then other dairy desserts. On average, diet yoghurts are sold at a higher price than the regular ones, whereas we observe the opposite for the cottage cheeses. This is explained by the fact that diet cottage cheeses are PLs only. Among NBs, there is some heterogeneity in the prices (Appendix Table 4.1). This heterogeneity might be linked to the heterogeneity in product characteristics but this also reflects strategic pricing by firms.

---

[8] The market share of a product (brand × retailer) is defined as the ratio of the sum of the quantities of the selected brand purchased at the selected retailer during a period of three months and the sum of quantities of all brands purchased at all of the retailers during the same period in the whole market, which includes dairy desserts, fruits, ice-creams, and pastries in the case of dairy desserts, and all kinds of milk in the case of fluid milk.

[9] One of the retailers only distributes 21 products rather than 23.

[10] The average is computed over the 159 products and the 16 periods (trimesters).

Price Transmission in Food Chains: The Dairy Industry

Table 4.1. Dairy desserts: Descriptive statistics for prices and market shares by categories

	Mean prices (€ per litre)			Market shares (%)		
	NBs	PLs	All	NBs	PLs	All
Regular yoghurts	2.33	1.45	1.90	23.0	21.7	44.7
Diet yoghurts	2.46	1.46	2.10	8.4	4.7	13.0
Regular cottage cheeses	3.05	2.00	2.51	7.9	8.2	16.1
Diet cottage cheeses		1.68	1.68		2.1	2.1
Regular forms of other dairy desserts	3.48	2.47	2.96	11.7	12.3	24.0
All products	2.72	1.81	2.27	50.9	49.1	100

All brands are available in every retail chain, except in retailer 7, which mainly sells PLs, which represent 80 per cent of its sales (Appendix Table 4.2). Retailers 1 to 5 choose similar prices. The average price in these retailers is mainly explained by the relative shares of NBs and PLs sold by each retailer. The higher the share of PLs sold by a retailer, the lower the average price in this retailer. Retailer 6 is between 10 per cent and 15 per cent more expensive than retailers 1 to 5, whereas retailer 7 is significantly cheaper. Retailer 7 is the aggregate of discounters, which generally offer fewer products with lower service levels. Indeed, prices for both NBs and PLs are lower in retailer 7 than they are in the other retailers.

4.2.3 The Market for Fluid Milk

In the case of fluid milk, the market is less diverse and there is almost no substitution between liquid milk and other beverages. For example, Bouamra et al. (2008) estimated a price elasticity of –0.15 for fluid milk in France. They also found non-significant cross-price elasticities between fluid milk and dairy desserts. In some studies, mostly examining the US market, fluid milk is a substitute for soft drinks. Allais et al. (2010), however, found that in France milk and soft drinks are not substitutes. Thus, econometric analysis suggests that the demand for liquid milk in France is highly inelastic and that substitution with other products is very limited. Therefore, in this analysis, we define the market as the fluid milk market. The bulk of the market comprises cow milk (98.6 per cent), with the remaining part being soy milk and goat milk. These two products are considered as the 'outside' option. Given the very small market share of these two products, it means that the market for cow milk will be considered as being inelastic. Relative price changes will mainly induce substitution between brands of liquid cow milk rather than substitutions with soy milk or goat milk. We define three segments for the cow milk market: skimmed, semi-skimmed, and whole milk. We select the main

Table 4.2. Fluid milk: Descriptive statistics for prices and market shares by categories

	Mean prices (€ per litre)			Market shares (%)		
	NBs	PLs	All	NBs	PLs	All
Skimmed milk	0.92	0.58	0.67	2.1	6.4	8.5
Semi-skimmed milk	0.71	0.63	0.66	34.1	51.7	85.8
Whole milk	1.07	0.73	0.84	2.0	3.8	5.8
All products	0.74	0.63	0.67	38.2	61.8	100

NBs. In our sample, Sodiaal and Lactalis are the two main manufacturers in the fluid milk market. Sodiaal has one brand (Candia), whereas Lactalis has two different brands (Lactel and Bridel). We also consider an aggregate NB produced by thirty-one smaller producers. For each of these four NBs, there are three products that are differentiated on the basis of their fat content. It should be acknowledged that for each brand, the market share of the semi-skimmed product is dominant. Considering the entire market, semi-skimmed milk has about 86 per cent of the market (Table 4.1). Finally, we define three PLs, one for each type of product. PLs dominate this market, with a market share as high as 62 per cent.

The average price over all products and all periods is €0.67 per litre (Table 4.2). The average price of whole milk is higher than that of skimmed and semi-skimmed milk, which have similar average prices. When differentiating PL and NB prices, however, the picture is a little bit different. Thus, for PLs, skimmed milk is cheaper than semi-skimmed milk, which is cheaper than whole milk. This is not true for NBs, as the price of semi-skimmed milk is significantly lower than the price of skimmed milk and whole milk. This is a clear indication of strategic pricing by firms. Due to the strong competition from PLs in the main market segment (semi-skimmed milk), NB prices are close to PL prices (NB prices are 'only' about 15 per cent higher). In respect of the other segments, the difference in prices between NBs and PLs is much higher (NBs are about 50 per cent more expensive than PLs). As shown in Appendix Table 4.3, among the NBs, there is some heterogeneity in prices. In particular, prices of the main NBs (brands 1 to 9) are higher than those of the aggregate NBs (brands 10 to 12). This certainly results from the market power and strategic pricing of the firms producing the main NBs.

As compared to the dairy desserts market, the market share of PLs in the fluid milk market is higher and the relative price difference between PLs and NBs is lower. A possible interpretation is that in the fluid milk market, consumers consider that the quality difference between PLs and NBs is not high, whereas they consider that there are still some differences in quality in the dairy desserts market. Then, in the fluid milk market, NBs cannot sustain higher prices (than PLs) and maintain significant market shares, whereas this is somewhat possible in the dairy desserts market.

All brands are available in every retail chain, except in retailer 2, in which a product is missing, and in retailer 7, which mainly sells PLs (Appendix Table 4.4). Retailers 1 to 5 choose prices that are quite similar on average. Despite the fact that PLs have almost the same prices in the different retailers, we note a larger difference between the prices of NBs. Retailer 6 and, to a lesser extent, retailer 3, sell the NB products at a higher price than the other retailers. Retailer 7 is still a special case in this market, with lower prices for both NBs and PLs than in the other retailers.

4.3 Models and Methods

To analyse strategic pricing in the food chain, we follow a general methodology that was recently developed to analyse vertical relationships between manufacturers and retailers (e.g., Berto Villas Boas, 2007; Bonnet and Dubois, 2010). We consider a demand model for each market to obtain the price elasticities of demand for every product in the market (that is, 159 products for the dairy desserts market and 103 products for the liquid milk market). The model needs to be as flexible as possible, so we opt for a random coefficients logit model (Berry, Levinsohn, and Pakes, 1995; McFadden and Train, 2000). Strategic pricing in the supply chain can be modified by the nature of the contracts between firms in the industry or by the vertical restraints considered. To deal with this issue, we design alternative models for the vertical relationships between the processors and the retailers. Based on the first-order conditions and estimates of demand, we are able to compute the price–cost margins for each product, from which we deduce the total marginal costs. To choose the vertical relationship model that best fits the data, we estimate a cost equation for each supply model and we use a non-nested Rivers and Vuong (2002) test to select the best one among all the alternatives.

Finally, using the selected model, we simulate the impact of a shock to the raw milk price on retail prices. In the following, we provide a brief summary regarding the main assumptions and methods. More extensive explanations regarding the details of the methodology can be found in Bonnet and Dubois (2010) and Bonnet and Réquillart (2013).

4.3.1 *The Demand Model: A Random Coefficients Logit Model*

We use a random coefficients logit model to estimate the demand model and the related price elasticities. The indirect utility function V_{ijt} for consumer i buying product j in period t is given by:

$$V_{ijt} = \beta_{b(j)} + \beta_{r(j)} + \alpha_i p_{jt} + \beta X_j + \varepsilon_{ijt}$$

where $\beta_{b(j)}$ and $\beta_{r(j)}$ are, respectively, brand and retailer fixed effects that capture the (time invariant) unobserved brand and retailer characteristics, p_{jt} is the price of product j in period t, α_i is the marginal disutility of the price for consumer i, X_j is a vector of product characteristics, and β is the vector of associated parameters. For dairy desserts, $X_j = (C_j, D_j, l_j)$ is composed of three dummies for cottage cheeses, dairy desserts, and diet products, respectively.[11] The associated coefficients represent the taste for cottage cheese and dairy desserts as compared to the taste for yoghurts and the taste for diet products as compared to the taste for regular products. For the fluid milk market, we chose two dummies for the fat category, $X_j = (S_j, W_j)$, for skimmed milk and whole milk, respectively.[12] The associated coefficients capture the taste for these fat contents as compared to the taste for semi-skimmed milk. Finally, ϵ_{ijt} is an unobserved individual error term.

In both markets, we assume that α_i varies across consumers. Indeed, consumers can have a different price disutility. We assume that the parameter has the following specification:

$$\alpha_i = \alpha_{NB_{j(i)}} + \alpha_{PL_{j(i)}} + \sigma \nu_i$$

where $\alpha_{NBj(i)}$ and $\alpha_{PLj(i)}$ are the mean price sensitivities when the product bought by the consumer is respectively a NB or a PL, ν_i captures the unobserved consumer's characteristics, and σ measures the dispersion of the unobserved heterogeneity from the mean price sensitivity. We assume a parametric distribution for ν_i denoted by $P_\nu(.)$ and P_ν is independently and identically distributed as a standard normal distribution.

We can then break down the indirect utility into a mean utility:

$$\delta_{jt} = \beta_{b(j)} + \beta_{r(j)} + (\alpha_{NB_{j(i)}} + \alpha_{PL_{j(i)}})p_{jt} + \beta X_j + \xi_{jt}$$

where ξ_{jt} captures all unobserved product characteristics and a deviation from this mean utility $\mu_{ijt} = p_{jt}\sigma\nu_i$. The indirect utility is then given by:

$$V_{ijt} = \delta_{jt} + \mu_{ijt} + \varepsilon_{ijt}.$$

The consumer can decide not to choose one of the considered products. Thus, we introduce an outside option that allows for substitution between the considered products and a substitute. The utility of the outside good is normalized to zero. The indirect utility of choosing the outside good is $V_{i0t} = \varepsilon_{i0t}$.

---

[11] Takes the value of 1 if product is a cottage cheese product and 0 otherwise; takes the value of 1 if product is another dairy dessert and 0 otherwise; takes the value of 1 if product is a diet product and 0 otherwise.

[12] Takes the value of 1 if product is a skimmed milk product and 0 otherwise; takes the value of 1 if product is whole milk product and 0 otherwise.

Assuming that ϵ_{ijt} is independently and identically distributed like an extreme value type I distribution, we are able to write the market share of product j at period t in the following way (Nevo, 2001):

$$s_{jt} = \int_{A_{jt}} \left(\frac{exp(\delta_{jt} + \mu_{ijt})}{1 + \sum_{k=1}^{J_j} exp(\delta_{kt} + \mu_{ikt})} \right) dP_\nu(\nu) \quad (1)$$

where A_{jt} is the set of consumers who have the highest utility for product j at period t, a consumer being defined by the vector $(\nu_i, \epsilon_{i0t}, \ldots, \epsilon_{ijt})$.

The random coefficients logit model generates a flexible pattern of substitutions between products, driven by the different consumer price disutilities α_i. Thus, the own- and cross-price elasticities of the market share s_{jt} can be written as:

$$\frac{\partial s_{ijt}}{\partial p_{kt}} \frac{p_{kt}}{s_{ijt}} = \begin{cases} -\dfrac{p_{jt}}{s_{jt}} \int \alpha_i s_{ijt} (1 - s_{ijt}) \varphi(\nu_i) d\nu_i & \text{if } j = k \\ \dfrac{p_{kt}}{s_{jt}} \int \alpha_i s_{ijt} s_{ikt} \varphi(\nu_i) d\nu_i & \text{otherwise.} \end{cases} \quad (2)$$

4.3.2 Identification and Estimation Method

For each market, an independent demand model is estimated using individual data. We randomly choose 100,000 observations among the 4,450,818 we have in the database of dairy desserts and also 100,000 observations among the 596,850 we have in the database of fluid milk.[13] We use the simulated maximum likelihood method, as in Revelt and Train (1998).[14]

This method relies on the assumption that in the dairy desserts market (the fluid milk market, respectively), all product characteristics $X_{jt} = (p_{jt}, C_j, D_j, l_j)$ ($X_{jt} = (p_{jt}, S_j, W_j)$, resp.) are independent of the error term ϵ_{ijt}. However, assuming that $\epsilon_{ijt} = \xi_{jt} + e_{ijt}$, where ξ_{jt} is a product-specific error term varying across periods and e_{ijt} is an individual specific error term, the independence assumption cannot be upheld if unobserved factors included in ξ_{jt} (and hence in ϵ_{ijt}) such as promotions, displays, or advertising are correlated with observed characteristics X_{jt}. For instance, we do not know how much each firm invests in advertising for its brands. This effect is thus included in the error term because advertising might play a role in households' choices of products. Because advertising represents an appreciable share of production costs, it is clearly correlated with prices. To address the problem that the omitted product characteristics might be correlated with prices, we use a two-stage residual

---

[13] Due to computing constraints, we were not able to estimate the demand model using the whole sample. The sample used is representative of the whole sample over products and periods.
[14] Models were estimated using 100 draws for the parametric distribution that represents the unobserved consumer characteristics.

inclusion method, as in Terza et al. (2008) and Petrin and Train (2010). The first stage consists of regressing prices on instrumental variables (input prices) and the exogenous variables of the demand equation brand and retailer fixed effects. This can be written as:

$$p_{jt} = W_{jt}\gamma + \delta_{b(j)} + \delta_{r(j)} + \tau X_j + \eta_{jt}$$

where W_{jt} is a vector of input price variables, γ is the vector of associated parameters, $\delta_{b(j)}$ and $\delta_{r(j)}$ are brand and retailer fixed effects, τ is the vector of coefficients associated with the exogenous variables of the demand model, and η_{jt} is an error term that captures the remaining unobserved variations in prices. The estimated error term $\hat{\eta}_{jt}$ of the price equation includes some omitted variables such as variations in advertising and promotions or in-shelf displays that are not captured by the other exogenous variables of the demand equation and by the cost shifters. Introducing this term into the mean utility of consumers δ_{jt} allows for the capturing of unobserved product characteristics that may vary across time. Thanks to this second stage, prices are now uncorrelated with the new product-specific error term varying across periods $\zeta_{jt} = \xi_{jt} + \varepsilon_{jht} - \pi\hat{\eta}_{jt}$.

We then write:

$$\delta_{jt} = \beta_{b(j)} + \beta_{r(j)} + \alpha p_{jt} + \beta X_j + \zeta_{jt} + \pi\hat{\eta}^{jt}$$

where π is the estimated parameter associated with the estimated error term of the first stage.

In practice, we use the price indices for the main inputs used in the production of desserts or milk because it is unlikely that the input prices are correlated with unobserved determinants of demand for these products. In the case of dairy desserts, we use the quarterly price of raw milk and the quarterly price indices of wages, gasoline, aluminium, glass, and metal. In the case of fluid milk, we use only the quarterly price of raw milk and the quarterly price indices of wages, gasoline, and cardboard.[15] These variables are interacted with PL/NB dummies because we expect that the manufacturers obtain different prices from suppliers of raw materials according to what they produce. We also expect that certain characteristics of the inputs depend on the PL/NB characteristics of products.

4.3.3 Supply Models: Vertical Relationships between Processors and Retailers, Cost Specification and Selection of the Best Model

Contracts between manufacturers and retailers, the degree of competition, and market power within the industry under investigation can modify the

---

[15] These indices are taken from the French National Institute for Statistics and Economic Studies.

price transmission of cost shocks (Bettendorf and Verboven, 2000; Bonnet et al., 2013). In food chains, both upstream and downstream industries are highly concentrated. In such a context of a chain of oligopolies, linear contracts are not efficient because, due to double marginalization, the profit of the chain is not maximized. More complex contracts allow the whole industry to maximize the industry profit. In this paper, we consider linear pricing and a set of two-part tariff contracts in which the processors have all of the bargaining power. The general framework of the vertical relationships is described by the following game:

- Stage 1: manufacturers simultaneously propose 'take-it-or-leave-it' contracts to retailers. In the case of linear pricing, the contract simply consists of a set of wholesale prices because the manufacturers produce a set of different brands. In this case, manufacturers compete à la Bertrand-Nash as they compete on prices. With a two-part tariff, the contract includes a set of wholesale prices and fixed fees. Finally, in the case of resale price maintenance (RPM), the contract is composed of a set of wholesale prices, fixed fees, and consumer prices.
- Stage 2: retailers simultaneously accept or reject the offers, which are public information. If a retailer rejects one offer, it earns some profit from an 'outside option'. We consider two possibilities. In the first case, the outside option is exogenously set to a positive fixed value. In the second case, the outside option is determined endogenously and its amount is equal to the profit that a retailer gets from selling its own PLs.
- Stage 3: retailers set consumer prices and thus compete à la Bertrand-Nash.

Depending on the assumptions regarding contracts and the outside option of retailers, we specify seven different cases: linear pricing, and six cases of non-linear contracts. The six cases of non-linear pricing derive from the combination of three types of contracts proposed by manufacturers with the two possibilities for the outside option of retailers. The three non-linear contracts correspond to a two-part tariff without RPM and two possibilities for a two-part tariff with RPM. With RPM, we consider the two polar cases for price–cost margins: zero wholesale margins for NBs ($w_j - \mu_j=0$) or, alternatively, zero retail margins for NBs ($p_j - w_j - c_j =0$).

Thanks to the first-order conditions derived from the supply models and demand estimates, we are then able to compute estimated price–cost margins of manufacturers and retailers for each product and estimated total marginal costs (for a detailed presentation of the different cases, refer to Bonnet and Dubois (2010) and Bonnet and Réquillart (2013)).

Once the demand model is estimated, for each model of vertical interactions between manufacturers and retailers, the price–cost margins can be computed. As prices are known, we obtain estimated marginal costs $C_{jt}^h = p_{jt} - \Gamma_{jt}^h - \gamma_{jt}^h$ for each product j at period t for any supply model h, where $\Gamma_{jt}^h = w_{jt}^h - \mu_{jt}^h$ is the manufacturer's margin and $\gamma_{jt}^h = p_{jt}^h - w_{jt}^h - c_{jt}^h$ is the retailer's margin.

We specify the following model for the estimated marginal costs in the case of dairy desserts:

$$C_{jt}^h = \sum_{k=1}^{K} \lambda_k^h W_{jt}^k + w_{b(j)}^h + w_{r(j)}^h + \eta_{jt}^h$$

and in the case of fluid milk:

$$C_{jt}^h = \sum_{k=1}^{K} \lambda_k^h W_{jt}^k + w_{r(j)}^h + \eta_{jt}^h$$

where W_{jt} is a vector of inputs, $w_{b(j)}^h$ represents the brand fixed effects for model h, and $w_{r(j)}^h$ is the retailer fixed effect for model h. We suppose that $E(\eta_{jt}^h|W_{jt}') = 0$ to consistently identify and estimate $\lambda_k^h, w_{b(j)}^h$ and $w_{r(j)}^h$.[16] To be consistent with the economic theory, as in Gasmi et al. (1992), we impose the positivity of parameters λ_k^h and use a non-linear least squares method to estimate them. We use this cost function specification to test any pair of supply models C_{jt}^h and $C_{jt}^{h'}$ and we infer which model is statistically the best using a non-nested Rivers and Vuong (2002) test.[17]

4.3.4 Simulations

Using the estimated marginal costs from the preferred model of contracts in the vertical chain as well as the other estimated structural parameters from the demand estimation, we can simulate the impact on retail prices of a change in the raw milk price. We denote $C_t = (C_{1t},..,C_{jt},..C_{Jt})$ the vector of marginal costs for all products present at period t. To model the impact of a change in the raw milk price, we have to solve the following programme:

$$\min_{\{p_{jt}^*\}_{j=1,..J}} ||p_t^* - \Gamma_t(p_t^*) - \gamma_t(p_t^*) - \tilde{C}_t|| \tag{3}$$

where $||.||$ is the Euclidean norm in $\mathbb{R}^J$, γ_t and Γ_t correspond respectively to the retail and wholesale margins for the best supply model, and $\tilde{C}_t$ is the vector

---

[16] In the case of fluid milk, we do not introduce brand fixed effects as we do in the dairy desserts case. This is explained by the fact that heterogeneous costs exist between brands of desserts, whereas fluid milk is quite a homogeneous product and its marginal cost does not differ across brands.

[17] In Section 4.3.4, we refer to this model as the preferred model of contracts.

Price Transmission in Food Chains: The Dairy Industry

of marginal cost estimated using the new raw milk price. If λ_R^h is the impact of the raw milk price on total marginal cost of the best supply model h, then we have $\tilde{C}_t = C_t + \theta\lambda_R^h$ where θ is the magnitude of the shock on the raw milk price.

4.4 Results for Demand, Vertical Relationships, and Cost Pass-Through

4.4.1 Demand Results

Table 4.3 provides the results for the demand estimates in both markets. First of all, the coefficient of the error term in the price equations is positive and significant for each sector.[18] It means that the unobserved part explaining prices is positively correlated with the choice of the alternative in each market and justifies the need to control for the endogeneity problem (we provide results for the price equations in Appendix Table 4.5). The instrumental variables used are not weak and significantly affect prices (cf. F-test of input price indices in each market). Moreover, correlation between instruments is not high.

In order to obtain better demand models, we introduced some product heterogeneity into the price sensitivity. Heterogeneity taken into account is related to the choice of NB or PL products. On average, the price has a significant and negative impact on utility. In both markets, consumers are more sensitive to the price variations of PLs than of NBs, suggesting that consumers might have more loyalty with respect to NBs than to PLs. The case of the fluid milk market is interesting. NBs' market shares are relatively low and the price difference between NBs and PLs is small. In other words, NBs cannot maintain significant market shares and high prices. This suggests relatively low product differentiation between both types of products. In this context, we interpret the difference in price sensitivity of consumers as follows: NBs' actual consumers are the proportion of consumers who are brand 'addicts' and do not switch easily to PLs. On the other hand, consumers of PLs can switch to the NBs if the price of PLs increases (Gabrielsen and Sorgard (2007) developed a model with such behaviour). It is also interesting to note that regarding retailers' fixed effects, we get similar results in both markets as fixed effects reveal consumers' preferences for a given retailer as compared to a reference one, which is the aggregate of discounters. Our results mean that preferences for retail chains are not product specific but are linked

---

[18] In the case of fluid milk, we assume that the error term can be different depending on whether it is an NB or a PL product. This allows taking into account some unobserved information that affects PLs and NBs in different ways. For example, PLs are frequently put in more favourable positions on retailers' shelves than NBs.

Table 4.3. Results of the random coefficients logit model

Dairy desserts			Fluid milk		
Variables price (p_{jt})	Mean	StD 0.028 (0.000)	Variables price (p_{jt})	Mean	StD 1.388 (0.000)
× PL	−2.058 (0.000)		× PL		−8.607 (0.000)
× NB	−1.792 (0.000)		× NB		−2.945 (0.000)
Yoghurt (Y_j)	—		Skimmed (S_j)		−2.009 (0.000)
Cottage cheese (C_j)	−0.473 (0.000)		Whole (W_j)		−1.271 (0.000)
Other dairy desserts (D_j)	1.380 (0.001)	—	Semi-skimmed (SS_j)		
Diet (I_j)	−2.095 (0.000)				
Brand fixed effects			**Brand fixed effects**		
B1 (Y-R-NB)	−0.648 (0.001)		B1		3.904 (0.000)
B2 (Y-R-NB)	−2.371 (0.001)		B2		3.726 (0.000)
B3 (Y-D-NB)	−0.543 (0.001)		B3		1.826 (0.000)
B4 (Y-D-NB)	1.688 (0.001)		B4		3.533 (0.000)
B5 (D-R-NB)	1.699 (0.001)		PL		8.449 (0.000)
B6 (D-R-NB)	−1.510 (0.000)				
B7 (C-R-NB)	0.351 (0.000)				3.904 (0.000)
B8 (Y-R-NB)	−0.928 (0.001)				3.726 (0.000)
B9 (C-R-NB)	0.825 (0.000)				1.826 (0.000)
B10 (Y-R-NB)	−2.859 (0.000)				
B11 (C-R-NB)	−0.213 (0.000)				
B12 (Y-R-NB)	−1.623 (0.001)				
B13 (Y-D-NB)	−0.538 (0.001)				
B14 (D-R-NB)	−1.528 (0.000)				
B15 (Y-R-NB)	0.044 (0.001)				
B16 (Y-D-NB)	0.028 (0.001)				
B17 (C-R-NB)	−0.312 (0.000)				
B18 (D-R-NB)	1.136 (0.001)				
B19 (Y-R-PL)	−0.978 (0.000)				
B20 (Y-D-PL)	−0.475 (0.000)				

B21 (C-R-PL)	—			
B22 (C-D-PL)	—			
B23 (D-R-PL)	—			
Retailers' fixed effects			**Retailers' fixed effects**	
R1	0.336 (0.000)		R1	0.067 (0.000)
R2	0.236 (0.000)		R2	0.286 (0.000)
R3	−0.102 (0.000)		R3	−0.141 (0.000)
R4	0.917 (0.000)		R4	0.611 (0.000)
R5	−0.293 (0.000)		R5	−0.126 (0.000)
R6	0.690 (0.000)		R6	0.264 (0.000)
R7	—		R7	—
Error term ($\hat{\eta}_{jt}$)	1.608 (0.000)		Error term ($\hat{\eta}_{jt}$)	
			× PL	2.557 (0.000)
			× NB	0.852 (0.000)
Log Likelihood	−320,960		Log Likelihood	−364,884
Number of observations	100,000		Number of observations	100,000

Standard errors are in parentheses.

to general characteristics of the chain, such as the services offered or the number of cashiers.

Let us now comment in more detail on the results for dairy desserts. Consumers prefer 'other dairy desserts' to yoghurts (as the mean coefficient for 'other dairy desserts' is positive, cf. Table 4.3) and yoghurts to cottage cheese (as the mean coefficient for cottage cheese is negative). They also prefer regular products to diet products. We also introduced brand fixed effects that reveal the mean preference of consumers for products.[19] Using the structural demand estimates, we compute own- and cross-price elasticities for each differentiated product. The own-price elasticities of demand for a brand vary between -8.27 and -2.97, with an average value of -4.77 (Table 4.4). Price elasticities are heterogeneous and in general rather high (in absolute terms). This could be explained by the number of alternatives that are considered in this market. Demand for regular products is more elastic than demand for diet products. Indeed, the average own-price elasticity of demand for regular brands of desserts is -5.07 while it is -3.92 for diet brands. If we compare the demand for yoghurts, cottage cheeses, and other dairy desserts, we observe that the demand for other dairy desserts is more elastic than the demand for cottage cheeses and both demands are more elastic than the demand for yoghurts, with average own-price elasticities of demand of -6.33, -4.90, and -4.04, respectively.

In the fluid milk market, households have a significant preference for semi-skimmed milk since the mean coefficients of both skimmed and whole milk are negative. The brand fixed effects reveal that PLs give the highest utility to households with respect to the other products. This might be explained by the fact that in this market consumers are more sensitive to the level of prices than to the brand they consume. One reason could be that fluid milk is a quite homogeneous product. The own-price elasticities of demand for a brand vary between -6.56 and -1.79, with an average value of -3.01 (Table 4.4). On average, elasticities of demand for the different product categories are similar, as the own-price elasticities of whole, semi-skimmed, and skimmed milk products are 3.36, -3.12, and -2.80, respectively. To interpret these elasticities, it is important to have in mind that these are product elasticities.[20] The literature frequently reports lower price elasticities, but they are generally evaluated for a more aggregated market. For example, Jonas and Roosen

---

[19] The coefficients of brand fixed effects cannot be compared directly as brands differ by characteristics that are taken into account in the preferences for diet (as compared to regular) products or for the product category. For example, to compare the preferences among NB yoghurts, the brand fixed effect of diet brands has to be corrected by the coefficient attached to the diet taste.

[20] In this analysis, a product is the combination of a brand and the location of purchase. A product is in competition with a large number of alternatives, which explains the very elastic demand for each product.

Table 4.4. Average own-price elasticities between products

Dairy desserts			Fluid milk	
Brands	Characteristic	Own-price elasticities	Characteristic	Own-price elasticities
B1	Y-R-NB	−4.25 (0.31)	S-NB	−2.25 (0.29)
B2	Y-R-NB	−3.82 (0.29)	SS-NB	−2.21 (0.14)
B3	Y-D-NB	−3.72 (0.32)	W-NB	−2.43 (0.09)
B4	Y-D-NB	−4.81 (0.44)	S-NB	−2.41 (0.13)
B5	D-R-NB	−8.27 (0.84)	SS-NB	−2.24 (0.15)
B6	D-R-NB	−4.82 (0.53)	W-NB	−2.45 (0.10)
B7	C-R-NB	−5.73 (0.54)	S-NB	−2.37 (0.12)
B8	Y-R-NB	−4.10 (0.56)	SS-NB	−2.35 (0.12)
B9	C-R-NB	−5.79 (0.44)	W-NB	−2.38 (0.13)
B10	Y-R-NB	−3.22 (0.26)	S-NB	−1.81 (0.23)
B11	C-R-NB	−4.98 (0.23)	SS-NB	−1.79 (0.09)
B12	Y-R-NB	−4.63 (0.50)	W-NB	−2.27 (0.20)
B13	Y-D-NB	−3.95 (0.52)	S-PL	−5.47 (0.58)
B14	D-R-NB	−5.83 (0.86)	SS-PL	−5.16 (0.50)
B15	Y-R-NB	−5.37 (1.33)	W-PL	−6.56 (0.71)
B16	Y-D-NB	−4.50 (1.06)		
B17	C-R-NB	−5.40 (0.87)		
B18	D-R-NB	−7.53 (1.75)		
B19	Y-R-PL	−2.97 (0.16)		
B20	Y-D-PL	−3.03 (0.17)		
B21	C-R-PL	−4.11 (0.22)		
B22	C-D-PL	−3.53 (0.47)		
B23	D-R-PL	−5.19 (0.59)		

Dairy desserts: Y/C/D stands for Yoghurt/Cottage cheese/Other dairy desserts; R/D stands for Regular/Diet; NB/PL stands for National Brands/Private Labels.
Fluid milk: S/SS/W stands for Skimmed/Semi-skimmed/Whole milk.

(2008) reported own-price elasticities for conventional milk around unity in the German milk market. When considering a larger number of alternatives, reported results are in line with ours. Thus, Lopez and Lopez (2009) found own-price elasticities ranging from −1.9 to −2.4 for different brands, and Kinoshita et al. (2001) found elasticities ranging from −0.2 to −6.1 depending on the brand and the store, with an average own-price elasticity of −2.8.

4.4.2 Preferred Model, Price–Cost Margins, and Cost Estimates

Using the demand estimates, we compute the price–cost margins and the marginal costs for each supply model. On the basis of the Rivers and Vuong (2002) test, the best supply model for the dairy desserts market is model '3', whereas it is model '6' for the fluid milk market (Appendix Table 4.6). These results mean that manufacturers and retailers use two-part tariff contracts with RPM and the distribution margin is equal to zero (meaning that the consumer selling price is equal to the sum of the wholesale price and marginal costs) in both markets. As in the French soft drink market (Bonnet and Réquillart, 2013), the French bottled water sector (Bonnet and Dubois, 2010), or the

German coffee market (Bonnet et al., 2013), we find that manufacturers and retailers use two-part tariff contracts with RPM. Then, the share of profits within the chain depends on the fixed fees, which are not observed. The PLs, however, play a strategic role in the fluid milk market, which is not the case for the dairy desserts market. The large market share of PLs and the strong substitutability between PLs and NBs in the fluid milk market could explain this result. In addition, there are many small firms that produce fluid milk (in 2010, there were 134 firms in this industry in France). It is therefore easy for retailers to find alternative producers for their procurement to replace NBs if needed. Then, PLs can play a strategic role.

Marginal costs, which include processing and retailing costs, are product- and time-specific. In the dairy desserts market, marginal costs of yoghurts, cottage cheese, and other dairy desserts amount to €1.52/kg, €1.99/kg, and €2.78/kg respectively, on average (Appendix Table 4.7). For yoghurts, there are no cost differences between regular and diet products. For cottage cheese, a diet PL is less costly to produce than a regular PL. Finally, for other dairy desserts, only regular products are produced. For each category, the marginal cost of the PL product is significantly lower than that of NBs. In the fluid milk market, marginal costs of PLs tend to be slightly higher than those of NBs (Appendix Table 4.8).[21] This result suggests that PLs and NBs are not different and in such a context NB producers are more efficient than PL producers (an assumption that is frequently made in the literature, e.g., Berges-Sennou and Bouamra (2012)). Among retailers, marginal costs in retailer 7 are the lowest ones. This is consistent with the strategy of retailer 7, which offers only a few services to consumers.

The raw milk price plays a significant role in the formation of marginal costs of all products (Appendix Table 4.9). For yoghurts, the coefficient for regular products is 1.16 but not significantly different to 1, which is the expected value, as producing one kilogram of plain yoghurt requires one litre of milk (Meyer and Duteurtre, 1998). The coefficient for the 'diet' yoghurt is lower than 1, which is also expected, as the production of low-fat yoghurt requires less milk equivalent since part of the fat is extracted. For cottage cheese, the coefficient is larger than 1, as expected, since producing one kilogram of cottage cheese requires more than one litre of milk. We find, however, a relatively low coefficient since the technical coefficient of production is larger

---

[21] This excludes the marginal cost of brands 10 and 11, which are underestimated, almost certainly because these products are priced at rather low prices for NBs. Thus, brands 10 and 11 are an aggregate of NBs with very low market share. Most are produced by relatively small firms. In our analysis, we choose to aggregate these brands because we do not have sufficient information to consider them separately. It is likely that these firms do not have the same market power as compared to the main NBs. By considering them as an aggregate, we almost certainly overestimate their market power and thus the margins. This is why we find very low costs.

than 2. Finally, for dairy desserts, it is much more complex to draw a parallel with the technology as these products are very diverse. In any case, milk is not the only ingredient and it thus makes sense to find a coefficient that is lower than 1. For fluid milk, the coefficient for whole milk is very close to 1 (0.94), which is in line with the technology, as production of one litre of fluid milk requires one litre of raw milk. For non-whole milk (we did not differentiate in the cost function between semi-skimmed and skimmed milk), the coefficient is lower than 1, which is also in line with the technology. It is, however, a little too low, as to produce a litre of skimmed milk requires 0.7 litre of milk equivalent (Meyer and Duteurtre, 1998).[22]

4.4.3 Simulations

To evaluate how retail prices change in response to a change in the raw milk price in those markets, we test the impact of a 10 per cent decrease in the raw milk price.[23] Using the estimated marginal cost and cost specification, we are able to assess the impact of a change in the raw milk price on the marginal cost of the different products. Then, we recover the new equilibrium prices using (3).

4.4.3.1 RESULTS IN THE CASE OF DAIRY DESSERTS

On average, the 10 per cent decrease in the milk price causes a 1.91 per cent, 1.99 per cent, and 0.55 per cent reduction in the total marginal cost of yoghurts, cottage cheeses, and other dairy desserts, respectively (Table 4.5). The 10 per cent decrease in the milk price has a lower impact on the marginal costs of diet products, which are low fat and thus contain less milk equivalent, than on marginal costs of regular products. On average, marginal costs of regular products decrease by 1.69 per cent, whereas it is 1.47 per cent for diet products (Table 4.5).

As a consequence of the reduction in marginal costs, consumer prices decrease by 1.1 per cent, 1.3 per cent, and 0.3 per cent for yoghurts, cottage cheeses, and other dairy desserts, respectively. The percentage variation of the price of dairy desserts is very small as these products have a rather low level

---

[22] In the cost model, we did not differentiate between semi-skimmed and skimmed milk, as this led to unrealistic results. When differentiating the two products, we obtained a marginal cost of production for semi-skimmed milk that was much lower than that of skimmed milk. This result is not plausible as the milk equivalent coefficient for producing semi-skimmed milk has been higher than that of skimmed milk. The difficulty comes from estimating the price–cost margins. Because we obtained identical price–cost margins and because, since 2008, the price of skimmed milk has been higher than that of semi-skimmed milk, we thus obtained a higher cost for skimmed milk. This explains why we chose not to differentiate between the two products in the cost equation, even if this led us to underestimate the marginal cost of production of semi-skimmed milk.

[23] We obtain the same results simulating a 10% increase in the raw milk price.

Table 4.5. Impact of a 10% decrease in the raw milk price on retail prices

Dairy desserts

		Change in cost in %	Change in price in %	Pass-through $\Delta p/\Delta c$
		Mean (std)	Mean (std)	Mean (std)
Brand 1	Y-R-NB	−2.09 (0.26)	−1.13 (0.12)	0.82 (0.00)
Brand 2	Y-R-NB	−2.34 (0.33)	−1.34 (0.17)	0.86 (0.01)
Brand 3	Y-D-NB	−1.23 (0.23)	−0.58 (0.09)	0.73 (0.01)
Brand 4	Y-D-NB	−0.88 (0.17)	−0.39 (0.06)	0.63 (0.00)
Brand 5	D-R-NB	−0.35 (0.05)	−0.21 (0.03)	0.72 (0.02)
Brand 6	D-R-NB	−0.73 (0.13)	−0.28 (0.04)	0.55 (0.01)
Brand 7	C-R-NB	−1.58 (0.21)	−0.99 (0.12)	0.84 (0.00)
Brand 8	Y-R-NB	−2.27 (0.43)	−1.19 (0.18)	0.82 (0.00)
Brand 9	C-R-NB	−1.55 (0.20)	−0.98 (0.12)	0.84 (0.00)
Brand 10	Y-R-NB	−3.07 (0.46)	−1.59 (0.20)	0.86 (0.01)
Brand 11	C-R-NB	−1.82 (0.23)	−1.19 (0.15)	0.88 (0.01)
Brand 12	Y-R-NB	−1.80 (0.31)	−1.14 (0.16)	0.88 (0.01)
Brand 13	Y-D-NB	−1.16 (0.33)	−0.58 (0.11)	0.77 (0.02)
Brand 14	D-R-NB	−0.55 (0.11)	−0.30 (0.05)	0.72 (0.02)
Brand 15	Y-R-NB	−1.86 (1.37)	−1.05 (0.45)	0.87 (0.01)
Brand 16	Y-D-NB	−1.17 (0.86)	−0.52 (0.20)	0.73 (0.02)
Brand 17	C-R-NB	−1.75 (0.55)	−1.14 (0.26)	0.88 (0.01)
Brand 18	D-R-NB	−0.45 (0.23)	−0.23 (0.09)	0.67 (0.02)
Brand 19	Y-R-PL	−3.48 (0.46)	−2.23 (0.27)	0.98 (0.01)
Brand 20	Y-D-PL	−1.70 (0.25)	−1.08 (0.14)	0.97 (0.01)
Brand 21	C-R-PL	−2.51 (0.30)	−1.85 (0.21)	0.99 (0.01)
Brand 22	C-D-PL	−2.66 (0.49)	−1.83 (0.28)	0.98 (0.01)
Brand 23	D-R-PL	−0.67 (0.11)	−0.52 (0.08)	0.96 (0.01)

Fluid milk

		Change in cost in %	Change in price in %	Pass-through $\Delta p/\Delta c$
		Mean (std)	Mean (std)	Mean (std)
Brand 1	S-NB	−2.88 (2.92)	−1.45 (0.41)	1.12 (0.23)
Brand 2	SS-NB	−3.18 (0.64)	−1.69 (0.25)	1.09 (0.02)
Brand 3	W-NB	−4.48 (0.77)	−3.14 (0.42)	1.30 (0.04)
Brand 4	S-NB	−2.27 (0.52)	−1.28 (0.25)	1.04 (0.04)
Brand 5	SS-NB	−3.00 (0.75)	−1.58 (0.28)	1.05 (0.03)
Brand 6	W-NB	−4.05 (0.79)	−2.95 (0.40)	1.33 (0.08)
Brand 7	S-NB	−2.55 (0.51)	−1.44 (0.24)	1.06 (0.02)
Brand 8	SS-NB	−2.58 (0.43)	−1.44 (0.21)	1.06 (0.02)
Brand 9	W-NB	−5.04 (1.57)	−3.33 (0.59)	1.25 (0.03)
Brand 10	S-NB	−7.84 (2.81)	−2.71 (0.61)	1.22 (0.08)
Brand 11	SS-NB	−7.02 (1.30)	−2.71 (0.37)	1.22 (0.05)
Brand 12	W-NB	−6.79 (2.89)	−3.83 (0.72)	1.28 (0.05)
Brand 13	S-PL	−2.62 (0.44)	−2.02 (0.26)	0.97 (0.01)
Brand 14	SS-PL	−2.57 (0.35)	−1.99 (0.24)	0.97 (0.01)
Brand 15	W-PL	−4.13 (0.78)	−3.41 (0.52)	1.00 (0.01)

NB/PL stands for National Brands/Private Labels; R/D stands for Regular/DietNB/PL stands for National Brands/Private Labels; Y/C/D stands for Yoghurt/Cottage cheese/Dairy desserts Labels;S/SS/W stands for Skimmed/Semi-skimmed/Whole milk.

of milk content. The pass-through, which is measured by the ratio of the difference in retail prices to the difference in marginal costs, is lower than 1 for all brands. For yoghurts, the pass-through has an average value of 0.83. Therefore, if the marginal cost decreases by €1 cent/kg, the retail price decreases by an average of €0.83 cents/kg. The industry thus under-shifts the cost change.

The pass-through is brand specific and varies from 0.55 to 0.99. For the NBs, it seems to be unrelated either to the firm or to the kind of product, even if the pass-through rates of other dairy desserts are the lowest ones (0.64 on average). The pass-through for PLs is larger than that of NBs. It is almost 1, meaning that retailers pass on to consumers most of the cost changes, which is not the case for manufacturers. A consequence is that the percentage reduction in PL prices is larger than that of NB prices, because PLs are cheaper than NBs and because retailers transmit cost changes to consumers to a larger extent. Because the contract between manufacturers and retailers includes RPM, manufacturers decide the final prices for their products. They choose a pricing policy for the entire set of products, thereby internalizing the substitution among their own set of products.

4.4.3.2 RESULTS IN THE CASE OF FLUID MILK

On average, the 10 per cent decrease in the milk price causes a €1.77 cents/litre reduction (approximately 4.1 per cent) in the total marginal cost of fluid milk in general. The impact on marginal costs is lower for semi-skimmed and skimmed milk (€1.31 cents/litre, or about 3.7 per cent) than for whole milk (€2.65 cents/litre, or about 5.0 per cent). As discussed above, the coefficients of the milk price in the cost equation were lower for semi-skimmed and skimmed milk than for whole milk. In response to the marginal cost change, consumer prices decrease on average by 1.81 per cent, 1.92 per cent, and 3.36 per cent for skimmed, semi-skimmed, and whole milk, respectively. For semi-skimmed milk, the pass-through has an average value of 1.11. Therefore, if the marginal cost decreases by €1 cent/kg, the retail price decreases by an average of €1.11 cents/kg. The industry thus over-shifts the cost decrease.

The pass-through varies from 0.97 to 1.33. Among the NBs, the greatest rates of pass-through are obtained for whole milk. The pass-through rate for PLs is smaller than that of NBs and also close to 1 in this market. Unlike the dairy desserts case, the percentage price reduction for the PLs is nearly the same as it is for the NBs, even if PLs are cheaper than NBs. This is because the cost pass-through rate for PLs is lower than that for NBs.

4.4.3.3 SYNTHESIS OF RESULTS

To sum up the results of cost pass-through rates across both markets, we regress cost pass-through estimates on market characteristics (Table 4.6). We defined

Table 4.6. Regression of pass-through on cost shock variables and product characteristics

	Model 1	Model 2
Retailer 1	0.01 (0.00)***	
Retailer 2	0.03 (0.00)***	
Retailer 3	0.02 (0.00)***	
Retailer 4	0.02 (0.00)***	
Retailer 5	0.03 (0.00)***	
Retailer 6	0.02 (0.00)***	
Retailer 7	–	
Fluid milk manufacturers		
Manufacturer 1	−0.07 (0.00)***	
Manufacturer 2	−0.10 (0.00)***	
Manufacturer 3	–	
Dairy desserts manufacturers		
Manufacturer 4	−0.49 (0.00)***	
Manufacturer 5	−0.40 (0.00)***	
Manufacturer 6	−0.46 (0.00)***	
Manufacturer 7	−0.45 (0.00)***	
Private labels	−0.25 (0.00)***	
Private labels (fluid milk)	0.00 (0.00)	
Fluid milk market	0.54 (0.05)***	
Manufacturer market shares	−0.54 (0.05)***	
Retailer market shares	−0.28 (0.06)***	
Private label shares	−0.49 (0.13)***	
Private label shares for PL	0.34 (0.04)***	
Const	1.19 (0.00)	1.01 (0.04)***
Period fixed effects	Yes	Yes
R-square	0.78	0.60

Standard errors are in parentheses.

two models. Model 1 assesses the differences between markets, producers, and retailers. Markets are taken into account through manufacturers.[24] The results suggest that there exists more heterogeneity across manufacturers or markets than across retailers.

As discussed in the previous section, the pass-through rates for NBs in the fluid milk market are higher than the rate in the dairy desserts market. This explains why the fixed effects associated with manufacturers 4 to 7 are much lower than those associated with manufacturers 1 and 2. Little heterogeneity exists between retailers even if retailer 7, the aggregated hard discounter, is the one that transmits lower cost shocks on average. Retailers transmit cost shocks onto PL prices in a similar way in both markets since the coefficient of the variable 'Private labels (fluid milk)' is not significant. Table 4.6 also presents the results of model 2, in which we regress the rate of pass-through on the manufacturers' and retailers' market shares, on the share of PLs, and on the market dummy. As anticipated, we found that the pass-through rate is higher

---

[24] We cannot identify the market effect from the manufacturer effect as there is no manufacturer present in both markets. Thus, manufacturers 1 to 3 are present in the fluid milk market whereas manufacturers 4 to 7 are present in the dairy desserts market.

for fluid milk than for dairy desserts. Taking this into account, we found that manufacturers with large market shares transmit less of the cost change than smaller manufacturers. A similar result is found for retailers as well. We also found that the share of PLs in the market has an impact on pass-through. The impact on the pass-through for NBs is negative. The higher the PL share is in a market, the lower the pass-through for NBs will be. On the contrary, the impact on the pass-through for PLs is positive. The higher the share of PLs in a market, the higher the pass-through for PLs will be. It should be noted that if there is a market effect, it is difficult to know if this is related to some characteristics of the market, or to the contracts that are in place. We have shown that the contracts used in these two markets are different and, as shown by Bonnet et al. (2013), the type of contract has an impact on price transmission.[25] Pass-through might also be related to the value of the elasticities. It is worth mentioning that for NBs, elasticities are much higher (in absolute terms) in the dairy desserts market than in the fluid milk market. With a lower elasticity, a price change has less impact on demand, which might explain a higher pass-through.

4.5 Conclusion

This chapter has provided a general methodology for evaluating price transmission in vertically related markets. This method allows us to assess the effects of changes in input prices on consumer prices by considering the pricing strategies of manufacturers and retailers in the food supply chain. We analysed the impact on two dairy markets of a change in the raw milk price. Using recent developments in the empirical industrial organization literature, we estimated a flexible demand model, a random coefficients logit model, and several models for the vertical relationships within the industry. We have shown that the most likely supply model is the model in which the manufacturers and retailers use two-part tariff contracts with RPM. In the dairy desserts market, PLs play no role in manufacturer/retailer relationships, whereas in the fluid milk market, they do play a role. In the latter case, PLs have a rather large market share, and retailers can procure fluid milk from several small producers. Moreover, products are not strongly differentiated in this market. Then, retailers can use PLs as a strategic tool in their negotiations with brand manufacturers. On the contrary, in the dairy desserts market,

---

[25] We evaluated the pass-through in the fluid milk market assuming the contract between manufacturers and retailers is characterized by RPM without a strategic role for PLs. We obtain very different results, in particular for skimmed and semi-skimmed NBs. For these products, the pass-through is lower than 1 (about 0.8). This is another example of the impact of the type of contracts on price transmission.

manufacturers still possess significant market power due to the strength of their brands. This result is consistent with anecdotal evidence with respect to this specific industry because the firms use investments in advertising to build strong reputations. Using the model that best fits the data, we have simulated the impact on consumer prices of a decrease in the raw milk price, taking into consideration strategic choices of the agents. We have shown that the pass-through for NBs is greater than 1 in the fluid milk market, whereas it is lower than 1 in the dairy dessert market. Depending on the market, therefore, firms will over-shift or under-shift cost changes. It should be noted that in both markets, we found that PLs transmit the cost change (the pass-through is very close to 1 for all PLs, whatever the market considered). Whereas under-shifting a cost change is the rule in a perfect competition framework (Stern, 1987), under conditions of imperfect competition, under-shifting or over-shifting is possible (Stern, 1987; Delipalla and Keen, 1992; Anderson et al., 2001). The literature also shows that conditions for over-shifting of a cost change are associated with conditions related to the elasticity of the slope of the demand curves. Moreover, as shown by Bonnet et al. (2013), the type of contracts linking manufacturers and retailers might also play a role.

According to these results, any analysis of the impact of food price policies requires that the strategic pricing of firms be addressed. We cannot, however, easily extrapolate our results to other industries. Neither the type of contracts used by a specific industry nor the qualitative results (e.g., the over-shifting of cost changes) can be generalized. First, the structure of the upstream industry plays a role in the choice of contracts between manufacturers and retailers. In the specific case of food markets, the structure of the upstream industry varies significantly from a low level of concentration (e.g., the meat industry or the wine industry) to a high level of concentration (e.g., the processed cheese industry or the bottled water industry). Second, the strategic response depends on the curvature of the demand function, which is also market-specific. Finally, the type of contracts also affects cost pass-through. As a consequence, the empirical analysis of price transmission in a given industry requires that the vertical relationships of this particular industry and the substitution patterns of consumers be evaluated first.

In our analysis, because we have assumed that consumers react identically to a price decrease or to a price increase, cost decreases or cost increases are transmitted in the same manner. The possibility of asymmetric price transmission is abundant in the literature that uses time series analysis. The general finding is that positive cost shocks are transmitted at a faster rate than negative cost shocks (for a recent survey, refer to Frey and Manera, 2007). Our study focuses more specifically on what occurs at the equilibrium. The model that we employ is static, and we focus on the change in equilibrium rather than on the speed or the path of adjustment. In that case, the results of an asymmetric price response (from time series analysis) might be less clear. For example, in his study, Peltzman (2000: 486–7) stated that: 'The important result is that there is no evidence of any permanent effect of asymmetries on the long-run trend of output prices: none of the relevant coefficients differ from zero. These results imply that the asymmetries do ultimately disappear but that it takes longer than five or eight months for this to happen.'

A second limitation relies on the relative homogeneity in the strategic behaviour of the different retailers. In reality, it appears that some heterogeneity exists in the way in which retailers adjust their prices. This observation stems from the fact that in the models constructed, retailers have similar vertical arrangements with manufacturers, while this might not always be the case in practice.

A third difficulty is to assess the role of small NBs in markets. The way to consider these brands is to aggregate them into a single NB. This approach is necessary because we do not have information on the manufacturers of those brands. By doing so, however, we likely overestimate their market power, leading to underestimation of their production costs. Taking into better account the role of these small NB producers is another challenge for this kind of approach.

Acknowledgement

We thank Olivier de Mouzon for his assistance in programming. Any remaining errors are ours.

Appendices

Appendix Table 4.1. Dairy desserts: descriptive statistics for prices and market shares by brands

	Type	Number of retailers	Prices	Market shares
			Mean (std)	Mean in % (std)
National Brands		7	2.72 (0.08)	50.94 (1.31)
Brand 1	Y-R	7	2.42 (0.11)	6.83 (0.64)
Brand 2	Y-R	7	2.10 (0.07)	2.15 (0.19)
Brand 3	Y-D	7	2.07 (0.08)	1.74 (0.24)
Brand 4	Y-D	7	2.75 (0.11)	4.27 (0.27)
Brand 5	D-R	7	4.57 (0.19)	2.27 (0.44)
Brand 6	D-R	7	2.70 (0.21)	4.64 (0.41)
Brand 7	C-R	7	3.25 (0.19)	1.75 (0.18)
Brand 8	Y-R	7	2.25 (0.17)	7.03 (0.64)

(continued)

Appendix Table 4.1. *Continued*

	Type	Number of retailers	Prices	Market shares
			Mean (std)	Mean in % (std)
Brand 9	C-R	6	3.25 (0.12)	2.21 (0.29)
Brand 10	Y-R	6	1.81 (0.06)	2.18 (0.31)
Brand 11	C-R	7	2.78 (0.08)	2.64 (0.21)
Brand 12	Y-R	7	2.61 (0.13)	1.59 (0.23)
Brand 13	Y-D	7	2.22 (0.09)	1.20 (0.31)
Brand 14	D-R	7	3.10 (0.15)	1.50 (0.13)
Brand 15	Y-R	7	2.74 (0.07)	3.23 (0.23)
Brand 16	Y-D	7	2.28 (0.19)	1.15 (0.29)
Brand 17	C-R	7	2.99 (0.07)	1.30 (0.16)
Brand 18	D-R	7	4.03 (0.21)	3.26 (0.19)
Private Labels		7	1.81 (0.05)	49.06 (1.31)
Brand 19	Y-R	7	1.45 (0.03)	21.73 (0.57)
Brand 20	Y-D	7	1.46 (0.02)	4.64 (0.36)
Brand 21	C-R	7	2.00 (0.05)	8.23 (0.37)
Brand 22	C-D	7	1.68 (0.07)	2.14 (0.26)
Brand 23	D-R	7	2.47 (0.08)	12.32 (0.39)

Y stands for yoghurt, C for cottage cheese, D for other dairy desserts, R for regular, and D for diet.
Standard deviations are in parentheses.
Prices are in euros/kg.

Appendix Table 4.2. Dairy desserts: descriptive statistics for prices and market shares by retailers

	Number of brands		Share of PL		Price of NB	Price of PL	Price	Market shares
	NB		PL	%	Mean	Mean	Mean	Mean in %
Retailer 1	18		5	46.95 (2.27)	2.64 (0.09)	1.74 (0.05)	2.22 (0.04)	17.45 (0.49)
Retailer 2	18		5	53.48 (2.29)	2.76 (0.11)	1.85 (0.07)	2.27 (0.04)	13.60 (0.20)
Retailer 3	18		5	34.53 (2.23)	2.71 (0.09)	1.77 (0.06)	2.39 (0.05)	10.52 (0.24)
Retailer 4	18		5	37.69 (2.13)	2.81 (0.09)	1.84 (0.05)	2.44 (0.05)	20.94 (0.55)
Retailer 5	18		5	42.52 (1.32)	2.69 (0.10)	1.83 (0.05)	2.32 (0.06)	7.87 (0.27)
Retailer 6	18		5	38.22 (2.02)	3.02 (0.10)	2.03 (0.09)	2.64 (0.05)	12.41 (0.39)
Retailer 7	16		5	81.34 (3.33)	1.90 (0.09)	1.73 (0.05)	1.77 (0.05)	17.20 (0.75)

Standard deviations are in parentheses.
Prices are in euros/kg.

Appendix Table 4.3. Fluid milk: descriptive statistics for prices and market shares by brands

	Type	Number of retailers	Prices	Market shares
			Mean	Mean in %
National Brands		7	0.74 (0.02)	38.16 (3.43)
Brand 1	S	7	0.95 (0.17)	0.88 (0.14)
Brand 2	SS	7	0.83 (0.05)	8.94 (1.31)
Brand 3	W	7	1.13 (0.08)	0.72 (0.10)
Brand 4	S	7	1.07 (0.09)	0.82 (0.10)
Brand 5	SS	7	0.86 (0.05)	7.44 (1.25)
Brand 6	W	7	1.16 (0.10)	0.62 (0.07)
Brand 7	S	5	0.98 (0.07)	0.02 (0.02)
Brand 8	SS	7	0.94 (0.07)	0.41 (0.07)
Brand 9	W	7	1.01 (0.07)	0.17 (0.03)
Brand 10	S	7	0.54 (0.03)	0.35 (0.12)
Brand 11	SS	7	0.57 (0.01)	17.34 (1.97)
Brand 12	W	7	0.89 (0.06)	0.45 (0.06)
Private Labels		7	0.63 (0.03)	61.84 (3.43)
Brand 13	S	7	0.58 (0.04)	6.38 (0.30)
Brand 14	SS	7	0.63 (0.03)	51.66 (3.47)
Brand 15	W	7	0.73 (0.04)	3.80 (0.28)

S stands for skimmed, SS for semi-skimmed, W for whole milk.
Standard deviations are in parentheses.
Prices are in euros/kg.

Appendix Table 4.4. Fluid milk: descriptive statistics for prices and market shares by retailers

	Number of brands		Share of PL	Price of NB	Price of PL	Price	Market shares
	NB	PL	%	Mean	Mean	Mean	Mean in %
Retailer 1	12	3	47.44 (0.08)	0.69 (0.02)	0.63 (0.04)	0.66 (0.03)	16.41 (0.59)
Retailer 2	11	3	72.03 (0.04)	0.71 (0.02)	0.66 (0.05)	0.68 (0.04)	12.83 (0.34)
Retailer 3	12	3	60.78 (0.03)	0.79 (0.06)	0.65 (0.03)	0.71 (0.04)	9.91 (0.43)
Retailer 4	12	3	41.81 (0.07)	0.75 (0.03)	0.70 (0.06)	0.72 (0.04)	18.81 (0.62)
Retailer 5	12	3	41.02 (0.03)	0.68 (0.02)	0.66 (0.04)	0.67 (0.03)	8.14 (0.32)
Retailer 6	12	3	58.47 (0.03)	0.91 (0.05)	0.66 (0.05)	0.76 (0.05)	10.43 (0.37)
Retailer 7	11	3	91.74 (0.01)	0.59 (0.02)	0.57 (0.02)	0.57 (0.02)	23.47 (0.92)

Standard deviations are in parentheses.
Prices are in euros/kg.

Appendix Table 4.5. Results on price equation

Dairy desserts		Fluid milk	
Coefficient (Standard Error)		Coefficient (Standard Error)	
Cow milk	0.730*** (0.361)	Cow milk	0.751*** (0.248)
Cow milk × PL	−0.172 (0.775)	Cow milk × PL	−0.227 (0.528)
Wage	−0.067*** (0.016)	Wage	−0.080*** (0.008)
Wage × PL	0.085*** (0.033)	Wage × M2	−0.040*** (0.010)
Aluminium	0.001 (0.001)	Wage × M3	−0.081*** (0.011)
Aluminium × PL	−0.000 (0.003)	Wage × M4	−0.054*** (0.012)
Gazole	−0.002*** (0.001)	Gazole	−0.001*** (0.000)
Gazole × PL	0.002 (0.003)	Gazole × PL	−0.000 (0.001)
Glass	0.015*** (0.005)	Cardboard	−0.005*** (0.002)
Glass × PL	−0.005 (0.010)	Cardboard × PL	0.005 (0.005)
Metal	−0.025*** (0.009)		
Metal × PL	0.017 (0.019)		
Cottage cheese	−0.262 (2.694)	Skimmed milk	0.104*** (0.012)
Other dairy desserts	0.282 (2.694)	Whole milk	0.214*** (0.012)
Diet	−0.294*** (0.047)		
Product fixed effects	167.68*** (0.000)	Brand fixed effects	20.51*** (0.000)
Retailers fixed effects	87.37*** (0.000)	Retailers fixed effects	28.99***(0.000)
R-squared	0.984	R-squared	0.960
Number of observations	2,574	Number of observations	1,514

*** significant at 5%; M2, M3, and M4 stand for Manufacturer 2, 3, and 4.
Standard errors are in parentheses for all coefficients except for fixed effects where p-values are in parentheses.

Appendix Table 4.6. Statistics T_n of non-nested Rivers and Vuong tests

Desserts								Fluid Milk						
$H_1 \backslash H_2$	2	3	4	5	6	7		$H_1 \backslash H_2$	2	3	4	5	6	7
1	0.65	-2.59	-1.17	0.56	0.07	-1.76		1	-1.88	-1.88	-1.12	-1.88	-1.88	-1.61
2		-3.76	-4.86	-0.58	-0.65	-2.26		2		-16.57	2.88	-26.17	-18.78	2.82
3			2.04	4.44	4.15	1.97		3			2.88	2.53	-20.90	2.82
4				3.61	1.46	0.35		4				-2.88	-2.88	-2.92
5					-0.48	-2.47		5					-10.79	2.82
6						-1.33		6						2.82

Model 1 is linear pricing
Model 2 is two-part tariff with RPM and w=μ
Model 3 is two-part tariff with RPM and p-w-c=0
Model 4 is two-part tariff without RPM
Model 5 is two-part tariff with RPM, w=μ and private labels buyer power
Model 6 is two-part tariff with RPM, p-w-c=0 and private labels buyer power
Model 7 is two-part tariff without RPM and private labels buyer power

For the dessert market, consider row 3: all test statistics of the Rivers and Vuong test are higher than 1.96 which means that model 3 is better than models 4, 5, 6, or 7. In addition, the test statistic in column 3 is lower than −1.96, which means that models 1 and 2 are not preferred to model 3. As a consequence, model 3 is the preferred model.

For the fluid milk market, consider row 6: the test statistic of the Rivers and Vuong test is higher than 1.96, which means that model 6 is better than model 7. In addition, the test statistics in column 6 are lower than −1.96 (except for model 1), which means that models 2, 3, 4, and 5 are not preferred to model 6. As regards model 1, we consider that model 6 is preferred at a 10% threshold (the test statistic is lower than 1.64). As a consequence, model 6 is the preferred model.

Appendix Table 4.7. Margins for the preferred model (dairy desserts)

By brands			By retailers		
	Total margins	Total marginal costs		Total margins	Total marginal costs
Brands	in %	in euros	Retailers	in %	in euros
B1	33.48 (2.62)	1.59 (0.18)	R1	29.16 (6.66)	1.83 (0.73)
B2	33.38 (2.57)	1.43 (0.16)	R2	27.55 (6.59)	1.97 (0.76)
B3	34.43 (3.20)	1.37 (0.18)	R3	28.84 (6.45)	1.84 (0.72)
B4	29.74 (3.07)	1.90 (0.24)	R4	27.50 (6.92)	2.03 (0.88)
B5	15.61 (1.46)	3.92 (0.46)	R5	28.27 (6.45)	1.90 (0.74)
B6	29.72 (3.08)	1.91 (0.29)	R6	25.47 (6.05)	2.20 (0.87)
B7	24.99 (2.52)	2.41 (0.30)	R7	32.77 (10.14)	1.63 (0.90)
B8	35.13 (4.35)	1.50 (0.32)			
B9	24.64 (2.11)	2.45 (0.25)			
B10	39.65 (3.00)	1.09 (0.14)			
B11	25.53 (1.10)	2.07 (0.12)			
B12	28.03 (3.16)	1.87 (0.28)			
B13	33.11 (4.94)	1.49 (0.29)			
B14	22.38 (2.72)	2.54 (0.48)			
B15	26.82 (11.01)	2.28 (0.75)			
B16	31.45 (11.16)	1.79 (0.59)			
B17	24.91 (5.03)	2.29 (0.49)			
B18	18.77 (6.76)	3.49 (0.98)			
B19	34.63 (1.91)	0.95 (0.08)			
B20	34.20 (2.13)	0.97 (0.08)			
B21	25.13 (1.43)	1.51 (0.11)			
B22	29.81 (3.64)	1.21 (0.23)			
B23	19.96 (2.18)	2.05 (0.29)			

Standard deviations are in parentheses.

Appendix Table 4.8. Margins for the preferred model (fluid milk)

By brands			By retailers		
	Total margins	Total marginal costs		Total margins	Total marginal costs
Brands	in %	in euros	Retailers	in %	in euros
B1	50.65 (9.97)	0.45 (0.26)	R1	45.76 (15.79)	0.43 (0.17)
B2	50.69 (3.62)	0.43 (0.09)	R2	43.65 (17.16)	0.48 (0.22)
B3	45.77 (1.89)	0.54 (0.19)	R3	43.75 (15.64)	0.47 (0.21)
B4	45.62 (2.83)	0.53 (0.22)	R4	44.96 (15.54)	0.46 (0.17)
B5	49.12 (4.05)	0.45 (0.11)	R5	42.85 (15.07)	0.53 (0.19)
B6	44.56 (1.93)	0.58 (0.26)	R6	43.20 (14.52)	0.52 (0.16)
B7	46.45 (2.36)	0.30 (0.27)	R7	49.39 (19.97)	0.25 (0.21)
B8	46.82 (2.44)	0.51 (0.11)			
B9	46.26 (3.18)	0.46 (0.22)			
B10	69.28 (8.27)	0.19 (0.14)			
B11	67.72 (3.90)	0.19 (0.03)			
B12	52.95 (6.62)	0.44 (0.13)			
B13	20.23 (2.91)	0.51 (0.07)			
B14	19.90 (2.12)	0.52 (0.06)			
B15	16.87 (2.63)	0.66 (0.10)			

Standard deviations are in parentheses.

Appendix Table 4.9. Estimation of the marginal cost function (preferred model)

Desserts Coefficient		Fluid Milk Coefficient	
C_{jt}	(Standard error)	C_{jt}	(Standard error)
Wages	0.0057 (0.0004)	Wages	0.0015 (0.0001)
Plastic	0.0004 (0.0000)	Cardboard	0.0004 (0.0000)
Energy	0.0055 (0.0006)	Energy	0.0008 (0.0001)
Milk × D	0.4776 (0.0325)	Milk × Whole	0.9352 (0.0485)
Milk × C × R	1.3294 (0.1258)	Milk × No Whole	0.4621 (0.0555)
Milk × C × L	1.1082 (0.3491)		
Milk × Y × R	1.1644 (0.1733)		
Milk × Y × L	0.5801 (0.0605)		
Coefficients $w_{b(j)}^h$ and $w_{r(j)}^h$ not shown		Coefficients $w_{r(j)}^h$ not shown	
F test for $w_{b(j)}^h$ (p value)	564.10 (0.00)		
F test for $w_{r(j)}^h$ (p value)	98.62 (0.00)	F test for $w_{r(j)}^h$ (p value)	47.22 (0.00)

References

AFSSA (2009). Etude individuelle des consommations alimentaires 2 (INCA2) 2006–2007, p. 228. Accessed 6 June 2014, available at, http://www.anses.fr/sites/default/_les/documents/paser-ra-inca2.pdf.

Allais, O., P. Bertail, and V. Nichele (2010). The effects of a fat tax on French households' purchases: a nutritional approach. *American Journal of Agricultural Economics* 92 (1): 228–45.

Anderson, S. P., A. de Palma, and B. Kreider (2001). Tax incidence in differentiated product oligopoly. *Journal of Public Economics* 81(2): 173–92.

Berges-Sennou, F. and Z. Bouamra (2012). Is producing a private label counterproductive for a branded manufacturer? *European Review of Agricultural Economics* 39(2): 213–39.

Berry, S., J. Levinsohn, and A. Pakes (1995). Automobile prices in market equilibrium. *Econometrica* 63: 841–90.

Berto Villas Boas, S. (2007). Vertical relationships between manufacturers and retailers: inference with limited data. *Review of Economic Studies* 74: 625–52.

Bettendorf, L. and F. Verboven (2000). Incomplete transmission of coffee bean prices: evidence from the Netherlands. *European Review of Agricultural Economics* 27(1): 1–16.

Bonnet, C. and P. Dubois (2010). Inference on vertical contracts between manufacturers and retailers allowing for non-linear pricing and resale price maintenance. *Rand Journal of Economics* 41(1): 139–64.

Bonnet, C., P. Dubois, D. Klapper, and S. Villas Boas (2013). Empirical evidence on the role of nonlinear wholesale pricing and vertical restraints on cost pass-through. *Review of Economics and Statistics* 95(2): 500–15.

Bonnet, C. and V. Réquillart (2013). Impact of cost shocks on consumer prices in vertically related markets: the case of the French soft drink market. *American Journal of Agricultural Economics* 95(5): 1088–108.

Bouamra, Z., V. Réquillart, C. Soregaroli, and A. Trevisiol (2008). Demand for dairy products in the EU. *Food Policy* 33: 644–56.

Bouamra-Mechemache, Z., R. Jongeneel, and V. Réquillart (2008). Impact of a gradual increase in milk quotas on the EU dairy sector. *European Review of Agricultural Economics* 35(4): 461–91.

Bukeviciute, L., A. Dierx, and F. Ilzkovit (2009). The functioning of the food supply chain and its effect on food prices in the European Union. *European Economy*, Occasional Papers 47. Brussels.

Delipalla, S. and M. Keen (1992). The comparison between ad valorem and specific taxation under imperfect competition. *Journal of Public Economics* 49(3): 351–67.

Food Drink-Europe (2014). Data and trends of the European food and drink industry 2013–2014. Food Drink-Europe. Brussels.

Frey, G. and M. Manera (2007). Econometric models of asymmetric price transmission. *Journal of Economics Surveys* 21(2): 349–415.

Gabrielsen, T. S. and L. Sorgard (2007). Private labels, price rivalry, and public policy. *European Economic Review* 51(2): 403–24.

Gasmi, F., J. Laffont, and Q. Vuong (1992). Econometric analysis of collusive behavior in a soft-drink market. *Journal of Economics & Management Strategy* 12(1): 277–311.

Hassouneh, I., C. Holst, T. Serra, S. von Cramon-Taubadel, and J. Gil (2013). Overview of price transmission and reasons of different adjustments in different EU member states. Technical report, TRANSFOP project (KBBE-265601-4).

Hellerstein, R. (2008). Who bears the cost of a change in the exchange rate? Pass-through accounting for the case of beer. *Journal of International Economics* 76 (1): 14–32.

Jonas, A. and J. Roosen (2008). Demand for milk labels in Germany: organic milk, conventional brands, and retail labels. *Agribusiness* 24(2): 192–206.

Kim, D. and R. W. Cotterill (2008). Cost pass-through in differentiated product market: the case of US processed cheese. *Journal of Industrial Economics* 55(1): 32–48.

Kinoshita, J., N. Suzuki, T. Kawamura, Y. Watanabe, and H. M. Kaiser (2001). Estimating own and cross brand price elasticities, and price–cost margin ratios using store-level daily scanner data. *Agribusiness* 17(4): 515–25.

Lloyd, T., S. McCorriston, W. Morgan, and E. Zgovu (2013). European retail food price ination. *EuroChoices* 12: 37–44.

Lopez, E. and R. A. Lopez (2009). Demand for differentiated milk products: implications for price competition. *Agribusiness* 25(4): 453–65.

Loy, J. P., T. Holm, C. Steinhagen, and T. Glauben (2015). Cost pass-through in differentiated product markets: a disaggregated study for milk and butter. *European Review of Agricultural Economics*, 42: 441–71.

McCorriston, S. (2013). Competition in the food chain. TRANSFOP, Working Paper No. 11. from the Transparency of Food Pricing Project under the EC Seventh Framework Programme (Grant Agreement No. KBBE-265601-4-TRANSFOP). Brussels.

McFadden, D. and K. Train (2000). Mixed MNL models for discrete response. *Journal of Applied Econometrics* 15(5): 447–70.

Meyer, C. and G. Duteurtre (1998). Equivalents lait et rendements en produits laitiers: modes de calcul et utilisation. *Revue Elevage Méditéranneén Médecine Vétérinaire Pays Tropicaux* 51(3): 247–57.

Nakamura, E. and D. Zerom (2010). Accounting for incomplete pass-through. *Review of Economic Studies* 77(3): 1192–230.

Nevo, A. (2001). Measuring market power in the ready-to-eat cereal industry. *Econometrica*, 69(2): 307–42.

Peltzman, S. (2000). Prices rise faster than they fall. *Journal of Political Economy* 108(3): 466–502.

Petrin, A. and K. Train (2010). A control function approach to endogeneity in consumer choice models. *Journal of Marketing Research* 47(1): 3–13.

Revelt, D. and K. Train (1998). Mixed logit with repeated choices: households' choices of appliance efficiency level. *Review of Economics & Statistics* 80(4): 647–57.

Rivers, D. and Q. Vuong (2002). Model selection tests for nonlinear dynamic models. *The Econometrics Journal* 5(1): 1–39.

Steiner, R. (1993). The inverse association between the margins of manufacturers and retailers. *Review of Industrial Organization* 8: 717–40.

Stern, N. (1987). The effects of taxation, price control and government contracts in oligopoly and monopolistic competition. *Journal of Public Economics* 32(2): 133–58.

Terza, J., A. Basu, and P. Rathouz (2008). Two-stage residual inclusion estimation addressing endogeneity in health econometric modeling. *Journal of Health Economics* 27(3): 531–43.

5

Spatial and Temporal Retail Pricing on the German Beer Market

Jens-Peter Loy and Thomas Glauben

5.1 Introduction

Food retail pricing is a highly complex issue. Recent research has explored several dimensions of the problem by analysing detailed scanner data sets, which are increasingly available for many markets, first starting with the US food retail markets. This has allowed researchers to focus on issues of cost pass-through, pricing behaviour across retail chains and different outlets (e.g., hypermarkets and supermarkets), and between branded and private label products, where research has highlighted considerable heterogeneity in the dynamics of pricing at the retail stage of the food chain. However, studies in this area to date have generally ignored the spatial and temporal aspects of retail pricing behaviour, and while these issues have drawn attention in the theoretical literature, there have been few empirical studies of these phenomena, particularly in the context of the EU food sector. The analysis of spatial and temporal dimensions of food retail pricing behaviour may very well reveal significant welfare implications; in addition, it will improve our understanding of the food retail pricing complexity. To this end, we focus on the German beer market, which is an ideal market to address the spatial and temporal aspects of retail pricing, due in part to the availability of detailed scanner price data but also to the regional and brand coverage of retail prices covering the geography of the German beer market.

Specifically, German consumers show a strong affection for locally produced beer brands. The use of regional specifics of landscape, culture, or peoples' attitudes is widespread in the marketing of German beer (Zühlsdorf and Spiller, 2012). Beer is a top-ranked product with respect to consumers' association with regionally produced food categories (DLG, 2011) and marketing managers therefore make use of the regional specifics to advertise brands. Web pages, television commercials, and newspaper ads show well-known and appreciated characteristics of the region of origin to create a unique and favourable brand image. For example, the brand Jever always shows some quiet spots on beaches along the German North Sea, while the brand Flensburger uses typical Northern German landscapes. In addition to the spatial aspect, beer consumption fluctuates seasonally in relation to holidays, special events (e.g., the European Football Championship), or due to warm weather. Beer consumption is up to 75 per cent higher in the summer than in the winter period (Private Brauereien Deutschland e.V, 2011).

Therefore, the main research question posed by this chapter is to what extent German beer brands employ spatially and temporally differentiated retail pricing and promotional strategies. Only a few studies have addressed issues of beer retail pricing. Recent examples are Rojas (2008), Rojas and Petersen (2008), and Slade (2004), who investigate the role of market power and the impact of advertising on consumption on the US and UK beer markets; they find no evidence of collusive behaviour. Rojas and Petersen (2008) find predatory and cooperative effects for advertising. Culbertson and Bradford (1991) show that beer prices vary substantially across US states due to demand, excise taxes, exclusive territories, and transportation costs. We add to the existing literature by analysing spatial and temporal retail pricing and promotional strategies for the top ten Pilsener and wheat beer brands in Germany at the level of individual retailers. The pricing and promotional strategies consist of three features, namely the regular price level, the frequency of promotions, and the size of promotional discounts. For these features, we estimate the impact of the (regional) origin of brands and of temporal shifters in demand, and we control for brand-specific and retail chain-specific variations of pricing strategies. The data under study are weekly retail scanner data for Germany.

We proceed as follows. First, we describe the German beer market and provide an overview of the data. Second, we develop the theoretical basis for spatially and temporally differentiated retail pricing strategies. Third, we describe the data and explain the empirical model specification. Fourth, we present the estimation results for all characteristics of the spatial and temporal retail pricing and promotional strategies. Finally, we summarize our findings and draw some conclusions.

5.2 The German Beer Market

To illustrate the spatial and temporal dimensions of the beer retail pricing and promotional strategies on the German market, we present some samples of beer retail prices for two different brands to capture the effects of space and time on regular price levels, the frequency of price promotions, and the sizes of promotional discounts. In Figure 5.1, we graph a few random samples for retail prices of the brands Radeberger Pilsener and Beck's Pilsener at different locations. The ordering of the graphs reflects the distance between the retailer location (point of sale) and the location of the brewery (production site). The distance measured in kilometres is mentioned in the title. Figure 5.1 shows for both brands that the average or regular prices go up with the distance and that price promotions appear to be used more often and with higher discounts on the brand's home market.

Before we address these issues more comprehensively in theory and data, some stylized facts about the German beer market are presented. Colen and Swinnen (2011) conclude that Germany, along with the US, UK, Czech Republic, and Belgium, is one of the major 'beer drinking nations' in the world. Fifty-three per cent of the total alcohol consumption in Germany comes from drinking beer. With an annual per capita consumption of about 100 litres, beer accounts for almost one-seventh of the total per capita beverage consumption in the nation.

In a survey carried out in 2004 and 2005, El Cartel Media (2005) investigated consumer preferences for beer in Germany. Results show that brands have a strong position in their local (home) market. Every participant in the survey knows at least one local brand. For every second respondent, the favourite beer brand comes from the region. For seventy per cent of the respondents, the brand is more important for the product choice than the price. Thus, German consumers are highly loyal towards their favourite regional brand. The market shares of brands distributed nationwide show that almost all of the top ten brands are market leaders in their regional (home) market. We define home markets as the regions in which the main brands' brewing facilities are located; we further define regions by the federal states of Germany. For example, the brand Radeberger is brewed in Saxony, while the brand Jever is brewed in Bremen. Table 5.1 shows the rankings of the regional market shares for the top ten pilsener and the top ten wheat beer brands in Germany for the years 2000 and 2001. Many brands have the highest market shares on their home market and / or in close regions.[1] The shaded cells indicate the federal state in which the respective brand is produced (home market).

Consumers can purchase beer in a variety of outlets, for example in specialized beverage shops (SBS), gas stations, or traditional food retail markets. Hard discounters (e.g., Aldi, Lidl, Norma), cooperative discounters (e.g., Plus, Netto), small and big supermarkets (e.g., Edeka), and small, regional, and national

---

[1] A similar table for wheat beers is available from the authors upon request.

Spatial and Temporal Retail Pricing on the German Beer Market

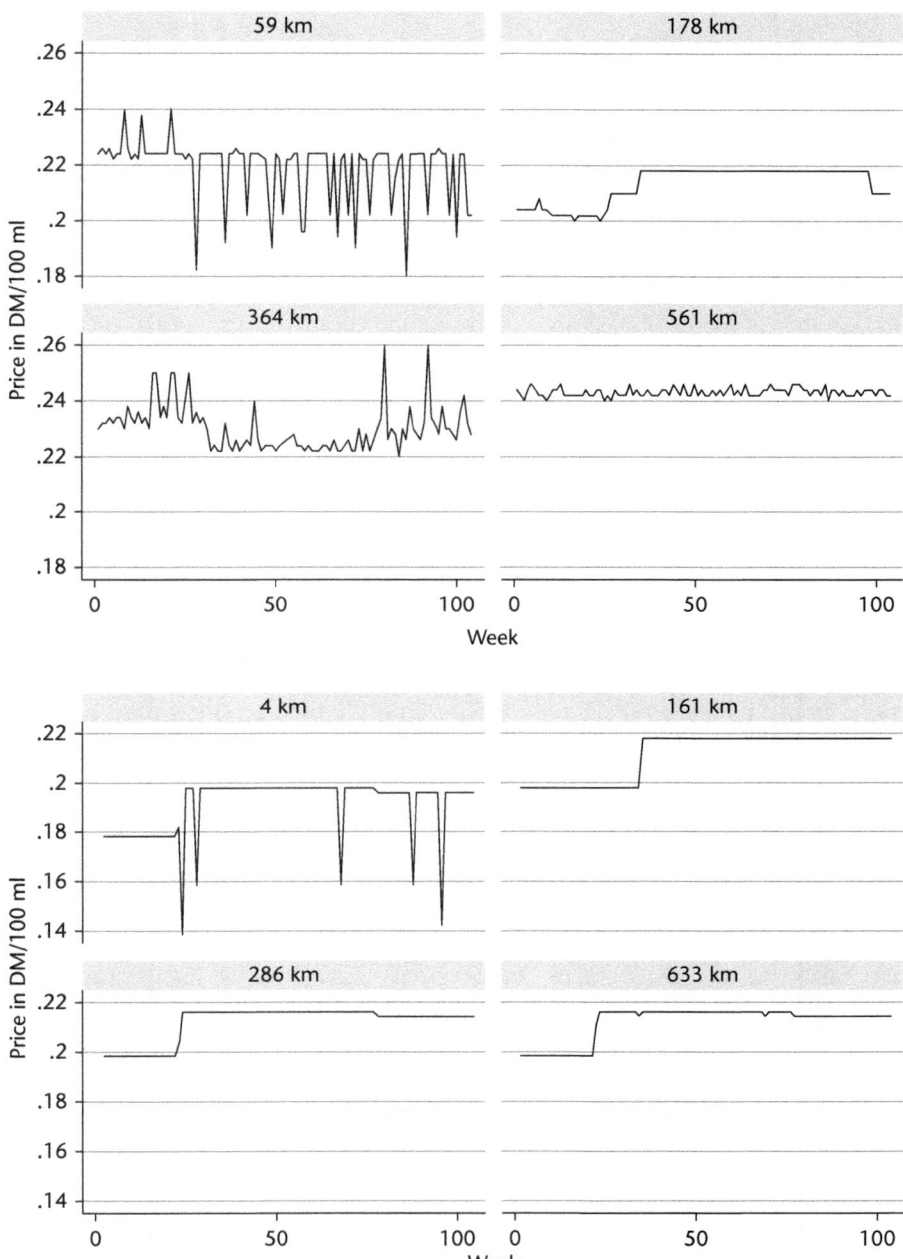

Figure 5.1. Sample retail prices of Radeberger pilsener for different distances between store and brewery (weekly from 2000 to 2001)
Source: Own based on MaDaKom, 2002.

Table 5.1. Ranks of regional market shares of beer brands in Germany (2000–01)

	Becks	Bitburger	Hasseröder	Holsten	Jever	Krombacher	Radeberger	Rothaus	Veltins	Warsteiner
Baden-Würt.	3	6	10	8	4	5	7	1	9	2
Bavaria	2	3	9	6	5	4	7	10	8	1
Berlin	2	8	5	4	7	6	3	10	9	1
Brandenburg	7	8	2	3	6	4	1	10	9	5
Bremen	1	7	9	3	5	4	8	10	6	2
Hamburg	4	6	8	1	3	5	8	8	7	2
Hessen	4	5	9	7	6	1	8	10	3	2
Mecklenburg Vorpommern	7	9	4	1	6	5	2	10	8	3
Lower Saxony	3	7	8	5	4	1	9	10	6	2
Northrhine-We.	5	3	9	7	6	1	8	10	2	4
Rhineland Pfalz	3	1	9	7	4	5	8	10	6	2
Saarland	5	1	7	9	4	3	6	10	8	2
Saxony	7	8	2	3	5	6	1	10	9	4
Saxony-Anhalt	8	9	1	2	6	4	3	10	7	5
Schleswig-Holstein	6	7	9	1	3	4	8	10	5	2
Thuringia	7	8	2	4	6	3	1	10	9	5
Mean Rank	5	6	6	4	5	4	6	9	7	3
Minium Rank	1	1	1	1	3	1	1	1	2	1
Maximum Rank	8	9	10	9	7	6	9	10	9	5
St. Dev.	2,2	2,7	3,2	2,6	1,2	1,6	3,1	2,3	2,2	1,4
Mean Market Share	7.7%	8.1%	5.8%	12.0%	5.0%	7.9%	7.9%	1.7%	3.2%	12.3%
Minium MS	0.4%	0.2%	0.0%	0.1%	0.5%	0.9%	0.0%	0.0%	0.0%	2.9%
Maximum MS	39.2%	47.2%	35.2%	66.7%	12.4%	21.5%	48.3%	26.5%	17.2%	20.7%
St. Dev.	9.5%	14.1%	9.5%	16.5%	3.3%	6.7%	12.6%	6.6%	4.7%	6.2%

Legend: Shaded fields mark the respective brand's home market.
Source: Own calculations based on MaDaKom, 2002.

hypermarkets (e.g., Famila, Plaza, Real) belong to the traditional food retail market in Germany. The traditional food retail market outlets account for about 50 per cent of the distribution of beer (Nahrung-Genuss-Gaststätten, 2009). The SBS make up 35 per cent of the market.

The top ten breweries produce 65 per cent of the total beer output in Germany of 100 million hectoliters per year. The top ten pilsener brands have a cumulative market share of more than 50 per cent. In the sample period, the top ten brands are Becks, Bitburger, Hasseroeder, Holsten, Jever, Krombacher, Radeberger, Rothaus, Veltins, and Warsteiner. Though the top ten brands have a significant market share, the German market is fragmented compared to the US or other international markets (Slade, 2004; *The Economist*, 2010). In 2000, the top four breweries in the US covered 95 per cent of the market, while in Germany the top four made up 30 per cent of the market (Adams, 2006). Seventy-five per cent of all European breweries are located in Germany. Import volumes are also less significant for the German market than for other European countries.

5.3 Theory

Traditional spatial and temporal pricing models consider the role of transportation and storage costs as well as the role of supply and demand fluctuations over space and time. Products transported over longer distances or stored over longer periods should be more expensive; also, when demand shifts upwards or downwards, prices should rise or fall. When supply expands or costs fall, the prices of goods should decrease. Varian (1980) challenged these comparative static results when he introduced the concept of a mixed strategy equilibrium. Following this concept, firms do not set deterministic prices based on supply and demand, but set prices according to a probability function. He shows that the probability function can have a u-shaped form, which favours high and low prices at the same time. This main result reflects the real-world phenomenon of price promotions or Hi-Lo pricing in retailing. For many food and beverage products, we observe price promotions implying that at the same time some retailers offer high regular prices and others offer significantly discounted prices. Many researchers have followed up on this concept and developed models for various more complex market settings. An important direction of research in this field of microeconomic modelling is investigating the role of consumer loyalty. There are a number of insights from the theoretical literature addressing a wide range of settings.

If consumers are mainly loyal towards brands produced in their own region, brands distributed nationwide by firms are likely to consider this in their spatial pricing and promotional strategies. Brand loyalty implies that consumers accept price differentials before they switch from their preferred to another competing brand. Not all consumers in the market may be loyal to the regional brand. Some may be non-loyal or switchers who react more sensitively to price changes. Thus, it may be rational even for strong (regional) brands to compete for switchers or non-loyal consumers by offering price promotions. For example, the regional brand may increase its market share by underbidding its competitors to gain all switchers' demand. This is not a profit maximizing strategy for all periods; however, a mixed strategy can be rational. Several papers have analysed the relationship between brand loyalty and promotional sales; the spatial aspect is to our knowledge not yet addressed in the literature. If we assume constant marginal costs of production for the brewers and restrictive transaction costs for consumers, we can solve the spatial pricing problem for each individual market separately and use the following models to obtain the impact of brand loyalty on (spatial) retail pricing.

Agrawal (1996), Anderson and Kumar (2007), Jing and Wen (2008), and Kocas and Bohlmann (2008) present models under various settings to determine the impact of brand loyalty on the retail pricing strategies of competitors. Agrawal (1996) models a retailer that sells a strong and a weak brand.[2] Both brands have loyal customers, but the stronger brand's loyal customers are willing to accept higher markups before switching to an alternative. The retailer faces two options: option one is to sell both brands at the consumers' reference price to the respective loyal segment. The second option is that the retailer offers a price promotion of either one of the brands to target the entire market. Because the level of loyalty is higher for the stronger brand, effective discounts for the weaker brand need to be higher. As this option is costly for the retailer (loss by price reduction in the loyal segment), it is used less often. Thus, local (strong) brands may promote more often but at smaller discounts compared to other non-regional brands. Jing and Wen (2008) use a different composition of consumer segments and introduce a non-loyal price-sensitive switching segment. They assume loyal consumers only for the stronger (local) brand and price sensitive consumers (switchers) for the others (non-local brands). Depending on the level of brand loyalty and the relative size of the respective consumer segments, different outcomes are possible. With a relative increasing price-sensitive consumer segment, brands offer deeper and more frequent promotions. Stronger brands will promote less aggressively when the degree of brand loyalty is high because it is more profitable to exploit the loyal segment than it is to attract price-sensitive consumers.

---

[2] The terms weak and strong indicate whether a brand has few or many loyal customers, or whether the loyalty is high or low (see Empen et al., 2014).

Kocas and Bohlmann (2008) introduce three brands and define the power of brands in relative terms. For brands with a larger segment of switchers and a smaller segment of loyal customers, it becomes profitable to offer higher discounts. Thus, stronger brands (more loyal customers) promote less often and offer smaller discounts.

Anderson and Kumar (2007) present a dynamic (two-period) model where brand loyalty is influenced by price promotions. Switchers choose the lowest-priced brand. For this brand, a fraction of switching consumers turn into loyal customers in the second period. Weak and strong brands indicate different rates of switchers that turn into loyal customers. Both brands face the trade-off between harvesting loyal customers by charging high regular prices and investing in a higher rate of loyal customers in the future. The stronger brand is more effective in turning switchers and thereby has a higher incentive to invest in future loyal customers. The strong brand therefore offers more and higher price discounts in the starting period.

In a standard model of perfect competition, positive (temporal) demand shifts lead to higher prices. Chevalier et al. (2003) discuss three different theories that deviate from the standard model. First, consumers are more engaged in shopping during times of high demand. Consumers are willing to search more intensively for low prices and demand becomes more price elastic. Consequently, retailers have more incentives to lower prices and to accept lower retail margins. Second, in periods of high demand, retailers rely less on tacit collusion. Costs of leaving a tacit cartel are equivalent to the sum of lower margins in future periods, of which currently higher market shares have to be subtracted. If there is a peak demand, revenues of deviating from tacit collusion increase. Consequently, more retailers lower their prices. Third, based on Lal and Matutes (1994), Chevalier et al. (2003) argue that retailers use price promotions to attract customers to the store. As advertising is costly, retailers cannot advertise all their goods. They choose products that are of particular importance for the store choice of consumers. The model by Lal and Matutes (1994) 'predicts it would be more efficient for retailers to advertise items that are relatively more popular' (Chevalier et al., 2003: 18). Therefore, retailers choose strong brands with a high share of loyal customers for price promotions.

If we follow a static concept of brand loyalty, then most models predict that stronger brands promote less often at lower discounts. For a dynamic concept, Anderson and Kumar (2007) show the reverse result, that stronger brands promote more intensively. This result can also occur following the loss leader theory when assuming that retailers choose primarily strong brands (key products) for promotion. Further, in periods of high demand, competition between retailers increases due to more intensively searching consumers.

5.4 Model Specification and Data

We analyse the pricing strategy of retailers based on the main three elements, namely the regular price level, the promotional frequency, and the average promotional discount (see also Bolton et al., 2007).

According to the theory, the composition of the consumer segment with regard to loyalty and the seasonal pattern of demand determine whether brands follow one of these strategies. In this chapter, we presume that the brands are strong in the region of origin, implying a high share of loyal consumers on the home market. Even though we do not directly measure the concept of brand loyalty, for example by consumer experiments or household scanner data, the market share of brands generally coincides with the level of brand loyalty (Fader and Schmittlein, 1993). We define the home market as the federal state in which the brewery operates its main production facilities. As shown in Table 5.1, most pilsener brands have a lead rank in the market share on their home market.[3] Seven out of the top ten brands rank first in their home market. Veltins is on rank 2 on their home market. Exceptions from this are the brands Jever and Warsteiner. Jever is nowhere on first rank, but shows similar rankings across Germany, likely due to its specific bitter taste; Warsteiner employed a marketing strategy that was strongly focused on the national market. Warsteiner ranked first or second in ten out of sixteen regional markets. Further, on the home market Warsteiner faces strong competitors, which rank first and second in this region (see Krombacher and Veltins in Table 5.1). Overall, Holsten and Warsteiner hold the highest market share, with 12 per cent each over the period of observation.

In the empirical model, we apply the distance of the retailer from the home market as an instrument to reflect changes in brand loyalty and/or transportation costs. To test whether the location of markets and seasonal demand shifts affect the brands' pricing strategies on the German beer market, we separately estimate the following model specification for all three dimensions of pricing strategies (PS: frequency of price promotion, relative size of the discount, or level of regular prices).

$$PS_{t,f,c,b,r} = \alpha + \beta Distance^{SP} + \mu Temperature + \sum_{i=1}^{7} \gamma_i D_i^{TH} + \sum_{i=1}^{2} \delta_i D_i^{FO}$$
$$+ \sum_{i=1}^{5} \epsilon_i D_i^{CH} + \sum_{i=1}^{19} \nu_i D_i^{BR} + \epsilon_{t,f,c,b,r} \qquad (1)$$

The dependent variables PS are the average share of promotional sales, the average relative discount, and the average regular price level for a particular brand (b) (out of the top ten pilsener and top ten wheat beers) over all stores of

---

[3] A similar table for wheat beer is available from the authors upon request.

the same format (f) and retail chain (c) in the same federal state (r) in the same week (t). To calculate the dependent variables, we need to identify promotional and regular prices. We follow Hosken and Reiffen (2001) and define sales as significant temporary price reductions that are unrelated to cost changes. More specifically, a sale indicates a price that is at least 5 per cent lower than the respective regular price of the brand during that period. A sale cannot last for more than four consecutive weeks; if a price cut lasts for more than four weeks we presume a regular price change. The regular or reference price is defined as the last non-sale price that persists for at least four consecutive weeks. For generating time series of regular prices, we replace the sales prices by the preceding regular prices respectively.

Following the theoretical considerations, spatial consumer preferences affect spatial pricing strategies. Thus, the main variable of interest is the 'distance' variable, measuring the distance from the location of the brewery to the federal state in which the retailer is located, respectively.[4] To consider deviations between brands, a set of dummies is added to the model (D_i^{BR}). We use retail chain (D_i^{CH}), retail format (D_i^{F}) dummy variables to consider firm- and format-specific differences, for example discounters usually follow an Every Day Low Pricing (EDLP) strategy with fewer price promotions and lower prices. Some retail chains may organize promotional activities nationwide, while others pursue decentralized promotional strategies, and so on.

Further, we introduce several time-dependent variables. The average daytime temperature for each federal state defines 'temperature' variable. Beer consumption, as well as the overall demand for beverages, depends on outside temperatures. To consider dynamic adjustments of the pricing strategy due to demand shifts during major holidays, we include another set of dummy variables (D_i^{TH}). We consider nationwide school holidays, Father's Day, Easter and Pentecost, Christmas, and New Year's Eve. In June and July 2000 the European Football Championship was held in Belgium and the Netherlands. This major event might have affected beer demand and/or marketing strategies. To capture this effect, a dummy 'UEFA Euro 2000' is introduced for the period of the tournament. Another time-dependent dummy variable is named 'Oktoberfest', which represents the period of the Oktoberfest for all retailers that are located in Bavaria.

The retail price data under study indicate a panel structure. Panel estimation techniques, therefore, may be more appropriate than quasi-fixed effects ordinary least square (OLS) estimates. Pretests show that a fixed effects panel model is preferred over a random effects model specification. However, as our main variable of interest is constant across panel members, we cannot use a fixed

---

[4] We take average distances between regional states.

effects panel model to deliver the respective estimates. Consequently, we decide to employ a quasi-fixed effects OLS model and a random effects panel model specification. Further testing shows that we need to correct for heteroscedastic and serially correlated errors; consequently, we adopt the estimation procedure outlined by Beck and Katz (1995) and use a two-step Prais-Winsten regression estimator to correct for serial autocorrelation. Further, we calculate robust standard errors to account for heteroscedasticity and cross-sectional correlation following the procedure by Chen et al. (2009).

We employ weekly retail scanner data provided by MaDaKom GmbH (2002) covering a two-year period from 2000 to 2001. The panel consists of more than 200 retail outlets. We select the ten top-ranked pilsener and the ten top-ranked wheat beer brands in the sample by calculating the overall average market share for the category pilsener and wheat beers. The top ten pilsener brands are Becks, Bitburger, Hasseroeder, Holsten, Jever, Krombacher, Radeberger, Rothaus, Veltins, and Warsteiner. The top ten wheat beers are Dinkel, Erdinger, Landskron, Loewenbraeu, Maisel, Oettinger, Paulaner, Scheider, Schoefferhofer, and Spatenbraeu. The data set includes all types of bottles and case sizes. We only use the most popular half-litre bottle size to exclude any effects related to the package (bottle) size. Retailer stores differ because they belong to certain retail (key) chains, for example Metro, Edeka, Rewe, Markant, and Tengelmann, and because their outlets show different formats, for example discounters (DC), small and large supermarkets (SM), and hypermarkets (HM).

In Table 5.2, the prices and sales measures under study are described for the various brands. Overall, brands' average regular prices range from 0.124 (Oettinger) to 0.258 DM (Warsteiner) per 100 ml. Sales frequencies range from 0 (Rothaus) to 5.03 (Radeberger) per cent. On average, every store puts beer on sale once a year. Average discounts range from 7.57 (Maisel) to 44.74 (Landskron) per cent; however, except for Landskron, most average discounts lie in a close range. Landskron only has a few observations and a very low rate of price promotion. The average discount is based on only two observations; thus, Landskron as well as Rothaus and Schoefferhofer do not use price promotions on a regular basis. Rothaus is a notable exception, offering no price promotions; Rothaus is the only beer brand in the sample that distributes exclusively on the home market Baden-Württemberg and follows an EDLP strategy by setting low prices with no price promotions. Most brands' average prices are in the range between 0.2 and 0.25 DM per 100 ml; however, for each brand we find significant deviations between average prices in specific retail outlets. For example, for the brand Becks we find retail stores that charge on average 0.19 DM per 100 ml over the sample period. The maximum average price of a retailer is 0.27. Many other brands show similar deviations. Thus, prices between brands seem to vary less than prices between retailers.

Table 5.2. Descriptive statistics of the brands' pricing strategies

Brand	% of Obs. (N)	Average regular price in DM/100 ml	Min. over stores	Max. over stores	Average freq. of sales in %	Average level of discount in %
Becks	10.1	0.221	0.191	0.270	1.87	9.93
Bitburger	6.6	0.229	0.200	0.298	2.38	10.50
Dinkel	1.1	0.209	0.197	0.260	1.21	10.09
Erdinger	8.4	0.246	0.197	0.317	4.83	10.29
Hasseroeder	3.8	0.209	0.187	0.257	2.40	9.63
Holsten	8.5	0.197	0.158	0.250	2.01	10.99
Jever	5.3	0.196	0.183	0.216	1.09	11.85
Krombacher	6.5	0.234	0.202	0.280	3.77	11.61
Landskron	0.3	0.179	0.161	0.197	0.49	44.74
Loewenbraeu	2.7	0.200	0.178	0.269	2.72	9.88
Maisel	1.0	0.252	0.229	0.267	8.12	7.57
Oettinger	2.3	0.124	0.107	0.187	2.08	22.58
Paulaner	11.0	0.245	0.172	0.298	4.74	11.57
Radeberger	6.1	0.230	0.182	0.294	5.03	10.11
Rothaus	0.5	0.229	0.224	0.232	0.00	0.00
Scheider	0.9	0.251	0.214	0.312	1.11	10.59
Schoefferhofer	3.7	0.220	0.198	0.261	0.68	12.54
Spatenbraeu	4.3	0.242	0.213	0.290	1.67	11.15
Veltins	4.7	0.200	0.194	0.264	1.13	10.49
Warsteiner	12.2	0.258	0.199	0.302	3.25	12.36
Overall (N)	126,665	0.225	0.107	0.317	2.92	11.11

5.5 Estimation Results

Table 5.3 presents descriptive statistics for the model variables that we use in the following estimations based on Equation 1. The dependent variables are described in more detail in Table 5.2. The main spatial variable is the distance between the brewery and the retailer location. The average distance is 340 km; the maximum distance is about 900 km. Thus, some breweries practice a nationwide distribution of their brands. The temporal variables are mostly holidays and the time during the European Football Championship in 2000. The average daily temperature over the sample period and regions is about 10 degrees Celsius, ranging from –5 to 25 degrees. Retail and chain dummies characterize the stores and key chains. Because hard discounters such as Aldi do not sell their scan data, the sample is biased. Besides the omission of hard discounters, the distribution of retailers in the sample reflects the number of different stores and store types. The distribution of brand dummies follows their sample observations.

We separately estimate the basic model specification shown in Equation 1 for all three dimensions of the pricing strategy, namely for the average mean

Table 5.3. Descriptive statistics of model variables (2000–01)

Variable	Description	Mean	Min	Max
Dependent Variables: PS				
Average Regular Price	Average of prices of the same brand within the same week, federal state, retail chain, and retail format	0.2256	0.1000	0.3200
Promotional Frequency	Share of promoted prices of one brand within one week, federal state, retail chain, and retail format	0.0292	0.0000	1.0000
Promotional Discount	Average share of percentage-based price reduction among discounted prices of a brand within one week, federal state, retail chain, and retail format	0.1111	0.0500	0.4747
Spatial Variable D_i^{SP}				
Distance	Distance between the retailer location (state) and the brewery location in 100 km	3.3977	0.02	8.91
Temporal Variables D_i^{TH}, D_i^{CH}				
Temperature	Average weekly temperature during daytime for the retailer location (state) in degrees Celsius	10.3556	−4.6858	24.5571
Father's Day	1 if the week preceding Father's Day, 0 otherwise	0.0195	0	1
Pentecost	1 if the week preceding Pentecost, 0 otherwise	0.0196	0	1
Easter	1 if the week preceding Easter, 0 otherwise	0.0193	0	1
Christmas	1 if X-mas (preceding and following week), 0 otherwise	0.0368	0	1
S. Holidays	1 if for nationwide school holidays, 0 otherwise	0.2488	0	1
Oktoberfest	1 during the Oktoberfest in Bavaria	0.0050	0	1
UEFA Euro 2000	1 if beer was sold during the European Football Championship in 2000	0.0264	0	1
Retailer Dummies D_i^{FO}				
Discounter	1 if beer was sold in a discounter, 0 otherwise	0.0816	0	1
Supermarket	1 if beer was sold in a retailer < 800 sqm not being a discounter, 0 otherwise	0.3625	0	1
Hypermarket	1 if beer was sold in a retailer >800 sqm not being a discounter, 0 otherwise	0.5559	0	1
Chain dummies	1 if beer was sold in a retailer not affiliated with a retail chain, 0 otherwise	0.0503	0	1
	1 if beer was sold in Edeka, 0 otherwise	0.2572	0	1
	1 if beer was sold in Markant, 0 otherwise	0.1518	0	1
	1 if beer was sold in Metro, 0 otherwise	0.2863	0	1
	1 if beer was sold in Rewe, 0 otherwise	0.0752	0	1
	1 if beer was sold in Tengelmann, 0 otherwise	0.1790	0	1
Brand Dummies D_i^{BR}				
Brand dummies	1 if beer was produced by Becks, 0 otherwise	0.1011	0	1
	1 if beer was produced by Bitburger, 0 otherwise	0.0663	0	1
	1 if beer was produced by Dinkel, 0 otherwise	0.0105	0	1
	1 if beer was produced by Erdinger, 0 otherwise	0.0841	0	1
	1 if beer was produced by Hasseroeder, 0 otherwise	0.0382	0	1

Spatial and Temporal Retail Pricing on the German Beer Market

1 if beer was produced by Holsten, 0 otherwise	0.0846	0	1
1 if beer was produced by Jever, 0 otherwise	0.0529	0	1
1 if beer was produced by Krombacher, 0 otherwise	0.0645	0	1
1 if beer was produced by Landskron, 0 otherwise	0.0032	0	1
1 if beer was produced by Loewenbraeu, 0 otherwise	0.0270	0	1
1 if beer was produced by Maisel, 0 otherwise	0.0103	0	1
1 if beer was produced by Oettinger, 0 otherwise	0.0231	0	1
1 if beer was produced by Paulaner, 0 otherwise	0.1096	0	1
1 if beer was produced by Radeberger, 0 otherwise	0.0614	0	1
1 if beer was produced by Rothaus, 0 otherwise	0.0054	0	1
1 if beer was produced by Scheider, 0 otherwise	0.0085	0	1
1 if beer was produced by Schoefferhofer, 0 otherwise	0.0374	0	1
1 if beer was produced by Spatenbraeu, 0 otherwise	0.0426	0	1
1 if beer was produced by Veltins, 0 otherwise	0.0469	0	1
1 if beer was produced by Warsteiner, 0 otherwise	0.1223	0	1

Source: Own calculation based on MaDaKom, 2002.

price, the share of promotional sales, and the average level of discounts.[5] To consider brand and outlet format-specific effects, we estimate a quasi-fixed effect panel model by introducing dummy variables for the different brands and outlet formats. Because of the different scales of the endogenous variables, we apply different estimation techniques. The frequency of sales is bound between 0 and 1; thus, a Probit quasi-fixed effect model is estimated. The share of promotional sales indicates a lower bound at zero. We therefore use a Tobit quasi-fixed effect model. The model for the average regular prices indicates heteroscedasticity, which we account for by calculating robust standard errors. In Table 5.4, we present the results for all models, which includes the results for the random effects panel model specifications. We predominantly rely on the results of the quasi-fixed effects model and random effects model. The results of the random effects model specifications are in brackets. Though the Hausman test results favour the fixed effects model, we present the random effects model to show the differences between brands and retailers.

---

[5] Error terms of all three equations might indicate some interaction. Application of a SUR estimation, however, is not necessary because the same exogenous variables enter the three equations.

Table 5.4. Estimation results for spatial and temporal beer brand pricing characteristics (2000–01)

	Average regional price (OLS-robust)	(Random effects model)	Average discount (Tobit)	(Tobit random effects model)	Average frequency (Probit)	Average frequency dF/dx (Probit)	(Probit random effects model)
Spatial effects: Base category is home markets							
Distance	0.0054***	(0.0041***)	−0.0055***	(−0.0021***)	−0.0204***	−0.0012***	(−0.0020+)
Temporal effects: Base category is all weeks in which none of these events took place							
Temperature	−0.0002***	(0.0000*)	−0.0022***	(−0.0002***)	0.0085***	0.0005***	(0.0103***)
Father's Day	−0.0027	(−0.0001)	0.0487***	(0.0024*)	0.1875***	0.0128***	(0.2299***)
Pentecost	0.0024	(0.0001)	−0.0384***	(0.0017)	0.1613***	0.0107***	(0.1917***)
Easter	−0.0042*	(−0.0004*)	0.0167	(−0.0004)	0.0582	0.0035	(0.0584)
Christmas	0.0015	(−0.0002)	−0.0528***	(−0.0005)	0.2083***	0.0143***	(0.2384***)
School Holidays	0.0019**	(0.0004***)	−0.0147***	(−0.0022***)	−0.0533**	−0.0030**	(−0.0610***)
UEFA Euro 2000	−0.0127***	(−0.0003)	0.0105	(0.0003)	0.0166	0.0010	(0.0282)
Oktoberfest	0.0168***	(0.0006)	0.0097***	(−0.0010)	−0.0421	−0.0023	(0.0370)
Retailer effects: Base category is discounters and retailers without affiliation to a retail chain							
Supermarket	0.0815***	(0.0892***)	0.0881***	(−0.0071***)	0.3089***	0.0193***	(0.6382***)
Hypermarket	0.0498***	(0.0590***)	0.0865***	(−0.0066***)	0.3056***	0.0170***	(0.6143***)
Edeka	0.0103***	(0.0131***)	0.0298**	(−0.0114***)	−0.1234**	−0.0066***	(−0.1567)
Markant	−0.0239***	(−0.0167***)	0.0481***	(−0.0114***)	−0.2084***	−0.0103***	(−0.1769)
Metro	−0.0557***	(−0.0602***)	−0.0308*	(−0.0062***)	0.1292***	0.0078**	(0.2021)
Rewe	0.0038**	(0.0148***)	0.0861***	(−0.0149***)	−0.3489***	−0.0149***	(−0.6808***)
Tengelmann	−0.0039**	(−0.0026)	0.0367***	(−0.0131***)	−0.1726***	−0.0088***	(−0.2311)
Constant	1.2287***	(1.230236***)	−0.5428***	(−0.0098)	−2.0817***	—	(−2.9181***)
N	124,805	(124,805)	125,448	(125,448)	124,763	124,763	(124,763)
R^2 [Pseudo]	0.6641	(0.9285)	[0.0661]	(0.0753)	[0.0487]	—	(0.0481)

Legend: +p<0.1, * p<0.05, ** p<0.01, *** p<0.001. Results of the panel specification in parentheses. The brand dummy estimates are not presented here, but can be obtained from the authors upon request.
Source: Own calculations based on MaDaKom, 2002.

Further, the main estimates of model parameters are similar for both specifications.

All specifications show high levels of overall significance. R-squares for the regular price equation range from 66 to 93 per cent.[6] The other models show moderate Pseudo-R-squares between 5 and 7 per cent.

The first row of Table 5.4 shows the estimates for the spatial dimension of the brands' beer pricing strategies. Overall, the coefficients for the distance variable are highly significant. The regular price increases and the measures of the promotional activity decrease with the distance from the home market. Thus, beer is on average cheaper in regular weeks (non-promotion periods) and brands promote more intensively on their respective home markets (more promotional sales at higher discounts). The coefficient of 0.0054 implies that the regular price for 100 ml of beer increases by 0.0054 DM (2 to 4 per cent with respect to average brand prices) every 100 km further away from the location of the brewery. The magnitude of the regular price effect appears to be economically significant considering that average regular prices of brands range between 0.124 to 0.258 DM per 100 ml. In view of the low percentage returns on sales in German retailing which are below 1 per cent, price differences of 2 or 4 per cent can be very relevant to business success. However, some of the spatial price effects can be due to costs of transportation.

The average discount level in the sample is about 11 per cent and the average frequency is about 2.9 per cent. The estimators for the distance variable suggest that brands almost offer no sales on distant markets, and that discount levels on these markets are 50 per cent lower compared with respective discounts on the home market. These results clearly support the theory by Anderson and Kumar (2007), who theoretically derive the result that the stronger brand is promoted more aggressively to generate new loyal customers.[7] As brands are stronger on their home market, price promotions are more frequently used. Thus, brands try to increase the number of loyal consumers by offering promotions. This strategy is most effective on the home market. The loss leader theory presents an alternative explanation. Retailers use price promotions to increase store traffic. Retailers use popular (key) products to effectively pursue this strategy. The products need to be relevant for the consumers' store choices. Beer, and in particular popular local brands, are of such a nature. Figure 5.1 displays the essential features of retail pricing strategies for the brand Radeberger over the sample period.

---

[6] The high R-square for the panel model is due to the consideration of first-order autocorrelation.
[7] Alternatively, the loss leader theory may present an explanation.

Temperature, dummies for national holidays (Easter, Ascension Day, Christmas, and Pentecost), and dummies for special events (UEFA Euro 2000, Oktoberfest) indicate the temporal differentiation of retail pricing strategies. Chevalier et al. (2003) show that prices fall during peak demand periods. The price calculated in the study by Chevalier et al. (2003) is a weighted average over all brands, including sales and regular prices. We analyse regular and sales prices separately to disentangle the substitution effects discussed in Nevo and Hatzitaskos (2006). For example, particular events such as Ascension Day or Pentecost, the effects of which last for a (few) day(s) only, may affect the promotional strategy but not the regular price level. Near Ascension Day promotional discounts are 0.05 per cent higher on average compared with other weeks. The likelihood of a promotion is also significantly higher for this event. In contrast, longer-lasting events such as the European Football Championship in 2000 may rather have an effect on the regular price, but not on the promotional strategy. The majority of coefficients indicate the expected signs: regular prices decrease and promotional activities increase during times of peak demand. Only school holidays deviate from this pattern: regular prices increase and the promotional activity slows down. The reason is likely to be product-specific, as school holidays do not indicate a close relationship to the consumption or demand of beer.

One of the most pronounced effects with respect to promotional discounts occurs for the week before Ascension Day. On average, discounts are 5.6 per cent higher than on other days. As the average discount equals 11 per cent, this represents an increase of 50 per cent. In Germany, Ascension Day or 'Father's Day' is a holiday on which many young men socialize outdoors while consuming alcoholic drinks, in particular beer. Ascension Day marks a peak in beer demand. During the European Football Championships the intensity of promotional measures is the same, but regular prices are set significantly lower in this period. Supermarkets and hypermarkets indicate higher regular prices of about 20 to 35 per cent compared to discounters. These retail formats apply a Hi-Lo strategy with a higher frequency of price promotions and higher price discounts. There is some evidence for chain-specific pricing effects. Warsteiner, the control group, shows the highest regular price level with a high intensity of price promotions (Hi-Lo). Oettinger is the cheapest brand that is following an EDLP strategy. The most promotions at the highest discounts are offered by the brand Maisel. The main results of the quasi-fixed effects model with respect to the distance or the temperature are very similar compared to the results of the panel model for all three specifications. Estimates of the frequency equation have the same sign but some are not statistically significant. The impact of temporal factors indicate most effects in the same direction across specifications, with differences confined to changes in significance.

5.6 Summary and Conclusions

The analysis of spatial and temporal retail price patterns can produce additional information on the conduct and the performance of markets. However, much of this information presently has no clear indication with respect to the impact on the society's welfare. For example, the finding that regular beer prices increase during the Oktoberfest may traditionally imply welfare losses for consumers. However, at the same time, price discounts are set at higher rates, implying better deals for (predominantly informed) consumers who may incur a potential extra cost to take advantage of the deal. Thus, uninformed consumers are mostly worse off and informed consumers with low transactions costs are likely better off. Further, if the latter group mainly shows low incomes, the second effect may outweigh the increase in regular prices due to income distributional aspects. This information is hidden when looking at average (national) prices. In addition, the choice in the procedure behind the averaging of prices may further induce an estimation bias, which further complicates a correct indication of welfare effects.

The results presented in this chapter clearly show an indication of spatial and temporal variations of pricing strategies over retailers and brands on the German beer market. According to several studies, local beer brands have more and stronger loyal consumers. The spatial pricing strategy is to promote more aggressively on the home market and to promote less on distant markets. However, on distant markets regular prices are higher than on the home market. For these features, we find two explanations respectively. Anderson and Kumar (2007) present a dynamic model of consumer loyalty to explain a more aggressive promotional strategy on the home market. Alternatively, Lal and Matutes (1994) have shown that loss leader pricing by retailers can produce a similar result. Depending on whether the retailer or the processor is in charge for setting retail prices, the one or the other explanation is more likely. The spatial pattern of regular prices may be the result of transportation costs or is part of a pricing strategy exploiting the niche markets. We also find evidence for temporal retail price patterns according to the theories and the data presented in Chevalier et al. (2003), who show that retail prices do not rise (fall) during periods of peak demand, for example before Christmas or Easter.

A managerial implication of these results may imply that expanding breweries on the German market are better off taking over competing (local) brands to conquer distant markets instead of heavily discounting their products (own brands) nationwide (on the distant markets). For takeovers on the German beer market in the past, we always find that new owners kept the brand's marketing concept of the newly acquired brand with almost no changes. Consequently, while brewers exit the market following a takeover, they leave their locally established brands' marketing concepts that have fostered consumer loyalty towards these brands.

The research reported here highlights certain limits that may call for a follow-up in future research activities in the field. First, though it is stated in the literature, the concept of beer consumers who are strongly loyal to local brands needs to be investigated by matching household scanner data with the retail scanner data to provide a direct measure of consumer loyalty. Empen et al. (2014) and Allender and Richards (2012) discuss and apply methods to match such data sources and present adequately derived measures of consumer loyalty. Thereby, we could further consider differences in consumer loyalty between brands. In this chapter, we argue that either loss leader pricing by the retailer or a dynamic pricing strategy by the processor could explain the spatial features in retail prices. To answer this debate, further information and an investigation of the role of processors and retailers in the setting of retail prices for the various brands are needed. Finally, many processors own several brands or sub-brands. To our knowledge, the pricing of such sub-brands has not been (theoretically) addressed to date.

References

Adams, J. W. (2006). Markets: beer in Germany and the United States. *Journal of Economic Perspectives* 20(1): 189–205.

Agrawal, D. (1996). Effects of brand loyalty on advertising and trade promotions: a game theoretic analysis with empirical evidence. *Marketing Science* 15(1): 86–108.

Allender, W. J. and T. J. Richards (2012). Brand loyalty and price promotion strategies: an empirical analysis. *Journal of Retailing* 88(3): 323–42.

Anderson, E. T. and N. Kumar (2007). Price competition with repeat, loyal buyers. *Quantitative Marketing and Economics* 5: 333–59.

Beck, N. and J. N. Katz (1995). What to do (and not to do) with time-series cross-section data. *American Political Science Review* 89(3): 634–47.

Bolton, R. N., V. Shankar, and D. Montoya (2007). Recent trends and emerging practices in retail pricing. In M. Kraft and M. Mantrala (eds), *Retailing in the 21st Century: Current and Future Trends*. 2nd edn. Heidelberg, Springer.

Chen, X., S. Lin, and W. R. Reed (2009). A Monte Carlo evaluation of the efficiency of the PCSE estimator. *Applied Economics Letters* 17(1): 7–10.

Chevalier, J. A., K. K. Anil, and P. E. Rossi (2003). Why don't prices rise during periods of peak demand? Evidence from scanner data. *American Economic Review* 93(1): 15–37.

Colen, L. and J. Swinnen (2011). Beer drinking nations. In J. Swinnen (ed.), *The Economics of Beer and Brewing*. Oxford: Oxford University Press.

Culbertson, W. P. and D. Bradford (1991). The price of beer. *International Journal of Industrial Organization* 9: 275–89.

DLG (Deutsche Landwirtschafts-Gesellschaft) eV. (2011). Regionalität aus Verbrauchersicht. Accessed 20 October 2012, available at, http://www.dlg.org/aktuelles_ernaehrung.html?detail/dlg.org/4/1/4479.

The Economist (2010). German beer drinking: Oktobergloom, producing too much, consuming too little: beer is a microcosm of Germany. 7 October 2010. Accessed 18 November 2011, available at <http://www.economist.com/node/17204871>.

El Cartel Media GmbH und Co. KG (2005). Branchenbericht Bier 2005, Grünwald. Accessed 12 October 2011, available at <http://www.elcartelmedia.de/xbcr/SID-FD7507F7-FFB0DAA5/ecm/Branchenbericht_Bier.pdf>.

Empen, J., J.-P. Loy, and C. R. Weiss (2014). Price promotion and brand loyalty: empirical evidence for the German ready-to-eat cereal market. *European Journal of Marketing*. Forthcoming.

Fader, P. and D. Schmittlein (1993). Excess behavioural loyalty for high-share brands: deviations from the Dirichlet model for repeat purchasing. *Journal of Marketing Research* 30(4): 478–93.

Gewerkschaft Nahrung-Genuss-Gaststätten (2009). *Branchenbericht der Brauwirtschaft 2009*. Hamburg, Ohne Verlag.

Hosken, D. and D. Reiffen (2001). Multiproduct retailers and the sale phenomenon. *Agribusiness* 17(1): 115–37.

Jing, B. and Z. Wen (2008). Finitely loyal customers, switchers, and equilibrium price promotions. *Journal of Economics & Management Strategy* 17(2): 683–707.

Kocas, C. and J. D. Bohlmann (2008). Segmented switchers and retailer pricing strategies. *Journal of Marketing* 72: 124–42.

Lal, R. and C. Matutes (1994). Retail pricing and advertising strategies. *The Journal of Business* 67(3): 345–70.

MaDaKom (2002). Markt Daten Kommunikation, retail scanner data set for 2000 and 2001. Duesseldorf.

Nevo, A. and K. Hatzitaskos (2006). Why does the average price fall during high demand periods? Working Paper, Department of Economics, Northwestern University.

Private Brauereien (2011). Absatzstatistik. Accessed 14 November 2011, available at <http://www.private-brauereien.de/pbb/presse-branche/absatzstatistik.php?navanchor=1010028>.

Rojas, C. (2008). Price competition in U.S. brewing. *The Journal of Industrial Economics* 56(1): 1–31.

Rojas, C. and E. B. Petersen (2008). Demand for differentiated products: price and advertising evidence from the U.S. beer market. *International Journal of Industrial Organization* 26: 288–306.

Slade, M. E. (2004). Market power and joint dominance in U.K. brewing. *The Journal of Industrial Economics* 52(1): 1–31.

Varian, H. R. (1980). A model of sales. *The American Economic Review* 70(4): 651–9.

Zühlsdorf, A. and A. Spiller (2012). Trends in der Lebensmittelvermarktung. Accessed 20 June 2012, available at <http://www.agrifood-consulting.de/index.php?id=cetest_firstpage3>.

6

The Use of Scanner Data for Measuring Food Inflation

Elena Castellari, Daniele Moro, Silvia Platoni, and Paolo Sckokai

> *Accurately measuring prices and their rate of change, inflation, is central to almost every economic issue. There is virtually no other issue that is endemic to every field of economics.* (Boskin et al., 1998)

6.1 Introduction

Changes in the cost of living are hard to measure in modern economies. First, markets are highly heterogeneous with a large variety of goods addressing the same need. Let's just think about the number of items available in a single store and the differences in assortment among stores: this will already lead to a number of items that are hard to handle in a simple price index formula. Second, prices vary quite dramatically over time. Promotion strategies are frequently used in modern retail markets, leading to frequent price changes. Similarly, the spatial competition among retailers leads to differences in prices for the same good. In a complex economy characterized by strong price dynamics and frequent changes in product characteristics, the calculation of an accurate inflation measurement becomes difficult.

Considering the complexity of the economy, the calculation of any price index requires some simplifying assumptions. Statistical agencies have to decide which commodities and products they want to use in their index calculations, where to collect data, at what frequency, and for how long they want to record the information needed. They need to establish criteria, consider the feasibility of their methodological choices, and make the index calculations as accurate as possible within resource constraints, This already explains the complexities hidden behind a simple index like the Consumer Price Index (CPI), traditionally used for measuring inflation.

The food sector is being recognized as one of the most critical for reaching an accurate measure of inflation, since food items play a crucial role in consumer expenditure and, at the same time, the food market is very complex. First, the food industry has a large fringe of small and medium enterprises operating in national or local markets. Second, especially after the recent developments of the retail industry, the food market is characterized by a large number of different items for the same type of product. Third, food prices are highly heterogeneous among different retail chains and types of stores. Fourth, food prices, as with energy prices, are recognized to be highly volatile, such that statistical agencies, when measuring the so-called 'core CPI', exclude the volatile prices of food and energy. Finally, the general change in food prices can have a considerable impact on the purchasing power of low-income households. In fact, when the food expenditure share of the household budget is high, even small price changes can lead to a trade-off with other needs. In this scenario, a more careful computation of food price inflation plays a crucial role in measuring the poverty status of households and, when necessary, in determining the need for or the amounts of assistance programmes. Moreover, government budgets as well as private contracts are tied to the changes in the CPI (Boskin and Jorgenson, 1997).

In 1996, a report by the Boskin commission estimated,[1] in the United States, an overall bias overstating the CPI of 1.1 per cent per annum. Specifically, the commission identified five potential sources of biases: product substitution bias, outlet substitution bias, new product bias, quality change bias, and elementary index bias. 'To attempt to account for these biases it has been acknowledged that both base period and current period price and quantity data is needed' (Ivancic, 2005).

Prior to our discussion, it is important to determine a common view on the role of the CPI and what we expect to measure with it. Traditionally, the CPI has been considered as a measure of the change in the cost of a fixed basket of goods and services, which are considered to be representative. However, especially after the Boskin commission report, statistical agencies moved from the simple connotation of the CPI as a price change measure to the interpretation of the CPI as a measure of the 'cost of living'. Measuring the cost of living compared to measuring the change in prices of a fixed basket of goods needs to allow the basket to change over time, in order to compare the minimum

---

[1] The Boskin commission, appointed by the US Senate in 1995, had the purpose of studying possible biases in the CPI computation and to propose recommendations of any needed change. The final report produced by the commission is titled 'Towards a most accurate measure of the cost of living'.

expenditure required to reach the same level of well-being considering different sets of prices (Boskin et al., 1996).

In this scenario, the use of high-frequency scanner data for measuring the cost of living can bring many advantages, especially when we refer to specific categories of products, such as food. First of all, one of the major advantages is due to the richness of the data being continuously collected. Moreover, when using bar-scan technology, there is no need of an *a priori* selection of the goods to be included in the basket and of the frequency in recording prices and quantities, since information is continuously collected for all items in stores. However, the challenges hidden behind the process of using this specific source of data are less explicit. For instance, the use of high-frequency data might not require that assumptions be made before the data collection, but it does not exempt statistical agencies from making decisions on the aggregation level at which the data can enter the CPI calculation. In particular, the type of aggregation over time, space, and items can become an important decision having a direct influence on the CPI volatility.

This chapter provides a review of the literature on the use of high-frequency scanner data for measuring the cost of living with a particular emphasis on food products. We highlight the major benefits and discuss the main challenges, such as: what are the effects of a time and spatial aggregation on the CPI calculation? What price index formula does the scientific literature advise for using scanner data?

In the second part of the chapter (Section 6.6) we investigate the effect of using high-frequency data for computing price indices by analysing scanner data collected at the store level for some dairy products in Italy. More specifically, we investigate the differences between the 'traditional' CPI computed by ISTAT and the corresponding indices computed with the scanner data, as well as the differences implied by the use of different index formulas on scanner data. However, before getting into the core of the chapter, in Section 6.2 we carry out a simple descriptive analysis of a sample of our scanner data, in order to get a feeling for their peculiarities.

6.2 The Dairy Scanner Data at a Glance

In this chapter we use a *Symphony IRI* data set recording weekly data on prices and quantities purchased of different dairy products at the item level, specifying whether the item is sold on promotion or not. This database covers 156 weeks, from January 2009 to January 2012, and 400 points of sales belonging to fourteen different retail chains. All points of sales in the sample are located in Italy, although the data set does not provide any information on their geographical location. For each point of sales we observe the retail chain it

The Use of Scanner Data for Measuring Food Inflation

belongs to and its store format (hypermarket, supermarket, or superette); discount stores are not included in the sample. Furthermore, in our data set the retail chain, the manufacturer, and the brand names are blinded by letter codes for confidentiality, even though we are able to identify private labels (PLs) and national brands (NBs). In this way we can distinguish different chains, manufacturers, and brands from one another, but we are not able to link them to real market entities. The data cover seven different dairy product categories: refrigerated and ultra-high temperature (UHT) liquid milk, butter, cheese,[2] mozzarella cheese, UHT cream, and yoghurt.

As is extensively discussed in the following sections, one of the peculiar features of using scanner data for price index computations is that such indices typically display a sizeable variability over time as compared to 'traditional' CPI computed by statistical agencies. Such variation is the result of both price and quantity changes that are in part due to the high level of disaggregation of the observed data, and in part linked to pricing and promotion strategies carried out by both manufacturers and retailers. Nakamura et al. (2011), analysing some retail products in the US market, found that much of the observed price variability takes place at the chain level. Similarly, exploring our data for two of the available products, high-quality refrigerated milk (Figure 6.1) and butter (Figure 6.2), we can observe both high price fluctuations within the same chain and differences in pricing and penetration strategies among chains.

Figure 6.1 shows the price trend along the 156 weeks for three NB products and the corresponding PL in chains A and B. The high-quality milk segment has on average a quantity share of more than 30 per cent of the entire refrigerated milk market in both chain A (31 per cent) and B (35.2 per cent).[3] While NB1 is the market leader in chain A, exhibiting an average market share of 44 per cent, in chain B it only has an average share of 5.3 per cent, thus showing a relevant difference in market penetration in the two chains. Similarly NB2, NB3, and the PL have higher market shares in Chain A (34.8 per cent, 18 per cent, and 10.1 per cent, respectively) with respect to Chain B (23.8 per cent, 11.7 per cent, and 8.5 per cent, respectively) even if this difference is not as pronounced as for NB1. Comparing the top and bottom panels in Figure 6.1, one can easily appreciate the differences in pricing strategies among brands. While NB2 exhibits a similar path between the two retail chains, NB1 and NB3 show different pricing behaviour, especially in the second half of the period. In chain A (top panel), NB2 reacts to a decrease in price by NB1 during late 2009–early 2010 with a strong reduction, while the NB3 price remains in line with NB1 for most of the period, although sales and

---

[2] Processed spreadable cheese and cheese flakes.
[3] The quantity market shares are computed as average of the weekly shares.

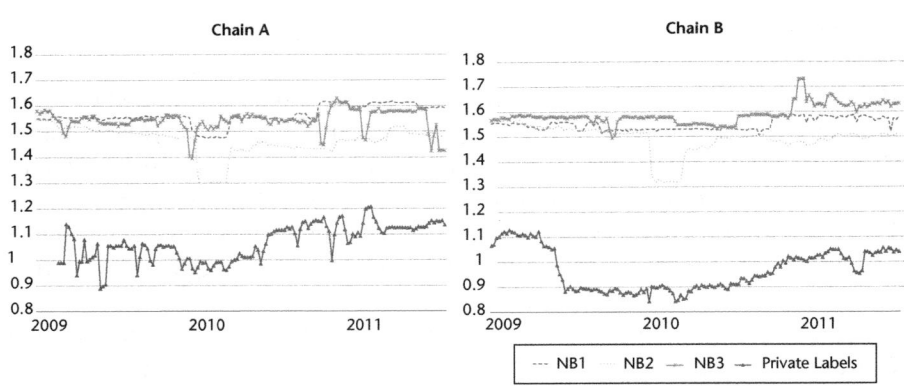

Figure 6.1. Brand level average weekly prices (euro/litre) for high-quality refrigerated milk in chain A (left panel) and chain B (right panel) (2009–11)
Source: Own elaboration on Symphony IRI data

promotional efforts are more frequent. On the contrary, in chain B (bottom panel), where NB2 is the market leader, its price reduction in late 2009–early 2010 does not induce any reaction by NB1, whose price tends to be rather stable, and always lower than NB3. Further, we observe an opposite pricing behaviour of NB3 during the year 2011: while the price path in chain A (top panel) shows several price reductions (sales/promotions), in chain B (bottom panel) we see a simultaneous upward tick at the beginning of the year with small fluctuations afterwards. Similarly, we notice that PLs display a consistently different pricing behaviour in the two chains. While in chain A (top panel) PL prices are characterized by sharp variations during the entire period, in chain B (bottom panel), after a sharp reduction in mid-2009, PL prices show a smooth path characterized by small variations. Most probably, PLs are exposed to higher promotion activities in chain A (top panel), even if overall the PL price is higher with respect to chain B. Differently, chain B (bottom panel) seems to carry out an 'everyday low price' strategy on their store brands.

Similarly, Figure 6.2 shows the price trend of three NBs and the corresponding PL for the same two chains (A top panel and B bottom panel) in the butter market. As we have previously observed for the high-quality milk segment, the market shares of different brands are highly different in chains A (NB1: 3.8 per cent; NB2: 2.8 per cent; NB3: 1.6 per cent) and B (NB1: 6.4 per cent; NB2: 4.7 per cent; NB3: 0.7 per cent), especially between PL products (6.7 per cent in chain A and 13.6 per cent in chain B). As before, brand level prices follow different paths. The graph for chain A (top panel) shows how the prices for the three NBs vary within a common narrow interval, while the NB2 price picks up in the second half of the period, remaining much higher than the other two NBs. Differently, in chain B (bottom panel), the graph shows a large difference

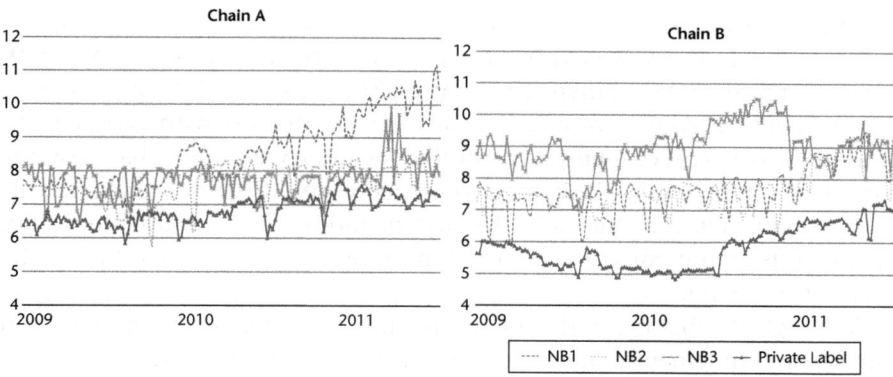

Figure 6.2. Brand level average weekly prices (euro/kg) for butter in chain A (left panel) and chain B (right panel) (2009–11)
Source: Own elaboration on Symphony IRI data

between the NB3 price and the other two NB prices, which becomes much smaller in the last twelve-month period, when the three NBs are sold at similar prices. Along the whole time period, PL prices are much closer to NB prices in chain A than in chain B, where overall the average PL price is much lower.

6.3 On Measuring the True 'Cost of Living'

The economic literature mainly presents two different approaches to estimate the cost of living indices (COI: Boskin et al., 1998). The first approach accounts for the substitution among different product categories using parameters of estimated demand systems. The second uses index formulas to compute the CPI. Usually, the calculation of the CPI is assessed through a pyramidal process where the all-item CPI summarizes all the variations of the consumer price indices in different commodity groups (energy, food, transportation, etc.), which in turn summarize the price variation of further disaggregated units within each category. The lowest level of aggregation is called 'item stratum' (Boskin et al., 1998).

While the first approach considers own and cross substitution among different product categories, the second approach assumes there is no substitution among products for a relative change in prices, meaning the cross-price elasticities are considered to be equal to zero. This strong assumption leads to bias in the cost of living measurement, since the CPI does not capture how consumers can adjust to relative price changes among different goods. For example, if we observe a 10 per cent increase of the relative price of apples with respect to pears, we expect many consumers to substitute apples with pears to adjust to the relative price change. The substitutability among products will not be captured by a traditional CPI computation, with a consequent bias of the real 'cost of living'. Specifically, the Laspeyres index,[4] based on the consumption pattern in the base period, usually tends to overestimate the true cost of living. On the other side, the Paasche index, which adjusts the cost of living considering current consumption levels, tends to underestimate the true cost of living.

Although the first methodology (estimation of a demand system) has been previously implemented for a fairly 'homogeneous' market, where products are considered at a high level of aggregation (Braithwait, 1980), its use becomes unfeasible when we refer to a lower level of product disaggregation. For instance, the number of parameters to be estimated rises almost with the square of the number of products. On the other hand, the second approach, besides the strong assumption on the absence of product substitutability, is much easier to implement, especially considering the strong product heterogeneity of a 'modern economy'. However, when we refer to the CPI as a measure of the 'cost of living' and not only as a change in the level of prices for a fixed basket of goods and services, we automatically introduce many other biases which are intrinsic with the data and the formula being used in a 'traditional' fixed basket approach. As mentioned in the introduction, the Boskin commission has categorized the different biases in five major sources:

1) *Substitution bias*: this refers to the inability of a fixed basket to account for substitution among different goods due to relative price changes (Boskin et al., 1996). There are different levels of substitution biases. The lower-level substitution bias occurs when the CPI fails to incorporate the shift in purchases among goods of the same type, that is, two brands of the same product, while the upper-level bias is due to incorrectly accounting for the change in expenditure among categories (Kliesen, 1997).

2) *Outlet substitution bias*: like the substitution among products, buying in one store instead of another can also have an important influence on the price paid. Since prices are collected within specific outlets, when consumers move from one shop to another the shift does not show up, leading to a bias (Boskin et al., 1998). Further, prices tend to be collected during the week, so all the weekend sales and price reductions will be omitted from the index computation.

3) *Change in quality bias*: over time, products change their quality. While estimating the cost of living, we would need to have quality-adjusted

---

[4] Please refer to Section 6.4 for a more comprehensive description of the indices.

prices, otherwise we may consider a quality improvement that is likely to increase consumer utility as a pure increase in price, thus generating inflation. This would misleadingly bias the CPI.

4) *New product bias*: when a new product is introduced in the market, the cost of living index is supposed to be adjusted considering the reservation price of the new item and the increase in surplus due to its introduction. However, when new products are introduced, the CPI accounts for their presence only after a relatively long time. Moreover, we can think of new products at different market levels, ranging from totally new categories of products (such as microwave ovens vs standard ovens), to different brands of the same product.

5) *Elementary index bias*: as explained before, the Laspeyres and Paasche indices are also biased by construction.

As Boskin et al. (1998) pointed out, the use of superlative price indices, such as the Fisher Ideal Index, can help reduce the bias generated by the Laspeyres/Paasche type of indices without introducing the dimensionality problem which will arise when using the demand equation methodology. The computation of superlative price indices requires the availability of more detailed data, and high-frequency data can be a good candidate for their computation. However, the use of such data may generate further biases that need to be carefully considered.

6.4 High-Frequency Data and CPI Computation

The use of high-frequency data can play a central role when trying to measure the cost of living using a CPI formula. In particular, scanner data are appealing for several reasons:

1) Scanner data record both price and quantity purchased. This information was previously unavailable to statistical agencies (Ivancic, 2005) and makes the calculation of superlative price indices more feasible, allowing a correction of the substitution bias carried out by the 'traditional' CPI. The opportunity to use a flexible basket index becomes a sizeable advantage when we refer to markets such as food, where products are highly differentiated, there is a high rate of entry and exit, and the assortment is strongly different among stores.

2) The use of the scanner technology allows for constant monitoring of the data with a precise and systematic procedure. By using high-frequency data we can observe the quantity sold and the actual price paid at the item level, thus at the most disaggregated level in time and space, while

in the 'traditional' procedure the price recorded is a shelf price, which usually is an average of product categories' prices that does not consider promotion activities. On the contrary, the use of scanner data allows us to record real purchasing consumer decisions, so the data implicitly account for the effects of marketing activities on consumer choices, allowing their incorporation when considering substitution patterns among different products.

3) There is no need to make a priori assumptions on the data collection procedure, leaving to a later stage the choice of the level of space and time aggregation for the CPI calculation. This can allow researchers to test the effect of different aggregation levels on the accuracy of the index calculation.

4) The number of goods recorded by scanner data is much larger than the number of goods included in the 'traditional' CPI fixed basket. This can have an effect in reducing the sampling error. For instance, Ivancic (2005) pointed out that even if scanner data are not able to provide a census for all goods, potentially they are able to increase the sample size for many goods with a consequent reduction of the sample variance (Bradley, 1996: cited in Ivancic, 2005; Richardson, 2000). The presence of a larger number of goods can have advantages in selecting a more representative basket for consumers. In addition, information on quantities can also help to provide more accurate weights to be used in the CPI computation (Reinsdorf, 1999: cited in Ivancic, 2005; Silver and Webb, 2002).

5) Scanner data give the possibility to monitor the expenditure shift among different stores. This gives the opportunity to make a more thoughtful selection of the outlets to be included in the index computation.

6) Further, when using scanner technology the introduction of new products is automatically considered, and this can help to reduce the 'new products' bias.

7) As already discussed, data collection for the calculation of the CPI can be highly demanding in markets where products are highly heterogeneous and prices tend to be very volatile. The use of scanner data for a large number of items might help in mitigating the need to implement manual collection procedures.

However, the use of scanner data in the CPI computation also has some technical and methodological drawbacks. The former relate to data management, while the latter relate to their usage in computing the CPI.

The first technical issue is related to the high computational burden implied by their use. This clearly implies higher costs for the statistical agencies for implementing a new procedure dealing with a much larger amount of data.

Moreover, at the moment, the cost of purchasing scanner data is rather high. However, their use can reduce the costs of direct data collection that are traditionally borne by the statistical agencies. Overall, it is necessary to evaluate additional costs and possible reductions on the current expenses to verify the budgetary impact of their introduction for statistical agencies (Ivancic, 2005).

Another technical issue is related to the time at which data are delivered, since statistical agencies are required to check and process such data within a very strict time schedule. The timing problem is part of a broader issue linked to the fact that statistical agencies would lose their direct control on the data collection process. Such loss may generate controversies and would require a strong partnership agreement between statistical agencies and data collectors (Bradley et al., 1997; Schut, 2001; Abraham, 2003).

On the methodological side, a first issue, as pointed out by Ivancic (2005), concerns the way in which scanner data are collected. It is true that scanner data contain a lot of information, however, when we observe a product disappearing from the data set, we do not recognize if this happened just because the sales of that item were zero in a given time period or because it was discontinued. Reinsdorf (1999) estimated that almost one-quarter of the products do not sell in a given week, so there is still an important share of products not represented in the scanner data set.

Another question is related to the interpretation of the CPI as a cost of living index that should measure the cost of the goods being consumed during a certain period of time. However, scanner data collect quantities purchased, and we cannot observe how many of these goods are being stocked up on and how many consumed. So, by measuring the CPI using scanner data, we intrinsically assume the goods bought have been immediately consumed in the given time period. Many authors have questioned this assumption (Bradley et al., 1997; Feenstra and Shapiro, 2001: as cited in Ivancic, 2005; Triplett, 2003). In particular Triplett (2003) argues that the discrepancy between purchases and consumption becomes larger if data are considered for a short period of time. This implies also that the level of time aggregation becomes important, since some biases may be related to the periodicity of purchases.

Further, even if the use of scanner data gives more flexibility during the data collection process, statistical agencies are required to decide how the data collected should enter the CPI computation. Specifically, one of the most important decisions is at what level the data should be aggregated. This issue is extremely relevant with the use of high-frequency data (Ivancic, 2005; Triplett, 2003). For instance, before starting to compute the CPI, users are required to decide:

1) The level of aggregation over time. Should we consider daily, weekly, or monthly prices and quantities purchased?

2) The level of aggregation over space. Are we going to consider prices and quantities at the store, retail chain, or territorial level?
3) The level of product aggregation. Do we analyse prices and quantities at the item, manufacturer, or product category level?
4) Do we let promotions and price reductions enter the CPI computation, or do we leave them out?

On the aggregation issue, researchers have been exploring whether data aggregation matters in the CPI calculation and, as described in Ivancic (2005), they have found significant differences among alternative aggregation schemes (Bradley et al., 1997; Dalen, 1997; de Haan and Opperdoes, 1997; Reinsdorf, 1999).

Moreover, if we try to compute price indices using high-frequency data with the conventional index formulas, we can obtain more volatile indices than using conventional data sources. In particular, the intensity of price and quantity bouncing due to sales has been found to cause chain drift bias when price indices are computed using high-frequency data (de Haan and van der Grient, 2011; Ivancic et al., 2009; Nakamura et al., 2011).[5]

The case of Norway is interesting in this regard. Since 2005 the Norwegian statistical institute has been using scanner data to compute the sub-index for food and non-alcoholic beverages, thus replacing the traditional procedure of calculating the food price index based on observations collected from selected items and sample stores. Changing the data source has brought much more variability, with a change in the products dimension from 250 items to 14,000 items. Rodriguez and Haraldsen (2006), comparing the price index computed using scanner data with the corresponding index computed over the traditional representative basket, showed how both sources predict the same index growth. However, while the indices for products such as milk, cheese, and eggs tend to be relatively stable using both procedures, other product categories, such as fruit, show a much higher volatility when the index is computed using high-frequency scanner data.

Nakamura et al. (2011) analysed several food categories and showed that specific aggregation over time and over stores can help reduce the price bouncing effect and consequently the chain drift bias in the index computation. In order to analyse the role of products, stores, and chains in explaining the cross-sectional variation in price dynamics, they implemented a variance decomposition model for three major food categories (coffee, soft drinks, and cold cereals). Specifically, they disentangled the variation common to all Universal Product Codes (UPCs) within a given product category from the

---

[5] The 'price and quantity bouncing' bias effect is linked to temporary sales, since households tend to stock up during sale periods and consume from inventory at times when the products are not on sale.

variation common to all items carrying the same UPC. Further, they decomposed the cross-sectional variation common across all stores, the one common only to stores within a chain, and the one related to items sold at specific stores. They repeated the estimation for the entire sample, either considering or not considering sales. The authors found that the source of variation is mostly derived from the frequency of temporary sales across chains for a given product rather than among stores of the same chain. Chain differences in pricing strategies dominate the price dynamics much more than store level differences, suggesting that statistical agencies should collect price data from a representative selection of retail chains. Further, they attributed the frequency of price changes across stores to several factors that turn out to be different across chains: the cost effectiveness of price decision making, the technologies used for implementing such price changes, the inventory management technologies, and the bargaining power toward the food industry (Nakamura et al., 2011).

However, as Ivancic (2005) pointed out, employing aggregation with the purpose of reducing the index volatility assumes that the 'observed volatility is "simply noise" and that volatility does not contain information that is meaningful in terms of trying to calculate a cost of living index'. On the contrary, index volatility may contain important information regarding consumer behaviour which might be valuable in measuring the cost of living (de Haan and van der Grient, 2011). Unfortunately, it is difficult to disentangle how much of the index volatility is due to important consumer behaviour information stored in the data and how much is the effect of the index formula that tends to inflate these biases.

A promising solution to reduce the volatility due to the chain drift bias is to resort to a drift-free chain index such as the Gini-Elteto-Köves-Szulc (GEKS) index suggested by Ivancic et al. (2011), since it allows for the use of data at a more disaggregated level, keeping the important consumer behaviour information which would be cancelled out using highly aggregated data sets.

6.5 Which Price Index?

There are several different numeric formulas to compute the CPI and they can be classified in different ways. A first classification is typically based on the fact that the basket of goods entering the index computation is either held constant (fixed basket index) or not (flexible basket index) over time. A flexible basket approach allows one to account for either the introduction of new products or a change in their characteristics. Another classification refers to the way in which the base period is treated in the index computation. While a chain index updates the base period over time, when using a standard price index the base period is held constant.

Laspeyres and Paasche indices are the most common price indices used for computing the CPI. Being p_{i0} the base period price for item i, p_{it} its price at time t for $t = 1, \ldots, T$, and w_{it} the good i's share of total expenditure at time t, the fixed basket Laspeyres index can be written as follows:

$$Laspeyres_t = \sum_i w_{i0} \cdot \frac{p_{it}}{p_{i0}} \qquad (1)$$

while its counterpart, the fixed basket Paasche index, can be written as:

$$Paasche_t = \left[\sum_i w_{it} \cdot \frac{p_{i0}}{p_{it}}\right]^{-1} \qquad (2)$$

A Laspeyres index measures the cost of a fixed basket of goods with respect to the cost in the base period. This index tends to overestimate the cost of living, since it does not allow substitution among goods and cannot consider the introduction of new products. Conversely, a Paasche index tends to understate the cost of living, since it weighs prices by current consumption patterns (Boskin et al.,1998; Diewert, 1998). Thus, when either the Laspeyres or Paasche indices are used, the resulting CPI cannot be considered a good measure of the cost of living.

As pointed out in the previous section, the use of superlative indices has been claimed to be preferable with respect to Laspeyres-type indices as they have been found to 'approximate the true cost-of-living index under certain assumptions'(Boskin et al., 1998) Moreover, the use of superlative indices can handle the potential dimensionality problem which can arise when estimating the cost of living through a demand equation approach (Boskin et al., 1998). The (unobservable) Pollak-Konüs true cost of living index has been found to fall between the Paasche and the Laspeyres price indices (Diewert, 1998). This result suggests that taking an average of Paasche and Laspeyres price indices can closely approximate the true cost of living. In particular the Fisher index, which is the geometric mean of the Paasche and Laspeyres indices ($Fisher_i = [Laspeyres_t \cdot Paasche_t]^{1/2}$), can be a good candidate to measure the cost of living (Diewert, 1998).

Specifically, Diewert (1976, 1995 and 1998) pointed out four main justifications for the choice of the Fisher index:

a) The base period basket used in the Laspeyres index is just as valid as the current period basket used in the Paasche index, thus it makes sense to take an even-handed average of the two.

b) The Fisher formula satisfies more reasonable 'tests' or 'axioms' than any of its competitors, including the reversibility property.

c) The Fisher is a 'superlative' index.

d) The Fisher index is consistent with revealed preference theory.

As discussed in the previous section, high-frequency data such as scanner data are suitable for calculating superlative price indices. However, as previous researchers have pointed out, the intensity of price and quantity changes due to sales can cause the so-called 'chain drift bias' (Ivancic et al., 2009; de Haan and van der Grient, 2011; Nakamura et al., 2011). For this reason, Ivancic et al. (2009) proposed the use of another index formula, the GEKS index, when trying to approximate the true cost of living using high-frequency data.

The GEKS index between i and k is the geometric mean of two Fisher indices (Ivancic et al., 2009). Consider P_{ij} to be the Fisher index between entities i and j ($j=1,\ldots,M$) and P_{kj} to be the Fisher index between entities k and j. We can, then, write the $GEKS_{i,k}$ index as follows:

$$GEKS_{i,k} = \prod_{j=1}^{M} \left[\frac{P_{ij}}{P_{kj}}\right]^{1/M} \qquad (3)$$

Ivancic et al. (2009) proposed to use the GEKS index to make comparison among T different time periods, $j = 1,\ldots,T$. Considering the reference time period $t = 0$, the GEKS price index between 0 and t, as in Ivancic et al. (2009), is the following:

$$GEKS_{0,t} = \prod_{t=0}^{T} \left[\frac{P_{0l}}{P_{lt}}\right]^{1/(T+1)} = \prod_{t=1}^{t} GEKS_{t-1,t} \qquad (4)$$

The circularity property of the GEKS index allows it to be written as a period-to-period chain index ($\prod_{t=1}^{t} GEKS_{t-1,t}$) between period 0 and period t ($GEKS_{0,t}$).

In addition, the GEKS index is free of chain drift bias as it satisfies the multi-period identity test proposed by Walsh (1901) and Szulc (1983). This property states that, given price indices among all different time periods, the price index will not suffer from chain drift bias if the product of indices among all possible time combinations is equal to one. For example, in the case of a three-time period, given the price indices between periods 1 and 2, $p(p_1,p_2,q_1,q_2)$, periods 2 and 3, $p(p_2,p_3,q_2,q_3)$, and periods 3 and 1, $p(p_3,p_1,q_3,q_1)$, if the product of the three indices is equal to one, the price index formula is not affected by chain drift bias. The GEKS index satisfies this multi-period identity property by construction.

Further, another advantage of the GEKS index is its suitability when using a flexible basket. Thus, the GEKS index is a good candidate for CPI computation using high-frequency data, since scanner data display a high heterogeneity in product assortment over time.

Despite these theoretical advantages, from a technical point of view the use of the GEKS index does not help in reducing the volume of the data. This can become a particular burden when data from a new time period become available, since all previous period values need to be recomputed, which of course implies a high cost for statistical agencies. Ivancic et al. (2009 and 2011) proposed the so-called Rolling-Window-GEKS (RWGEKS) index as an extension of the GEKS that adopts a moving window approach to update the price series as new data come in.[6] The use of the RWGEKS does not require the revision of the previous period computations, but only the extension of the calculations to the new period. Specifically, Ivancic et al. (2011) suggested the Rolling-Year-GEKS (RYGEKS) to statistical agencies. The RYGEKS is a RWGEKS with a 'window' of thirteen months, which allows the comparison of prices also for seasonal products, which may not be sold in all periods. However, statistical agencies should be careful in interpreting the RWGEKS indices, since the transitivity property is not satisfied among different windows but only within each window. Thus, among windows the index will be potentially subject to chain drift bias (Ivancic et al., 2011). Other approaches can be investigated, but at the actual state this is the best available proposal among superlative indices.

6.6 The Italian Dairy Market Case Study

In this section we discuss the peculiarities of computing price indices for food products using high-frequency scanner data. Figures 6.1 and 6.2 have shown the high variability observed in high-frequency data both in terms of price trends and in terms of market shares at the chain level. Of course, this price and share variability strongly affects the computation of the price indices. For this reason, we have used our scanner data set for two main purposes:

1) To compare the indices computed on scanner data with the official CPIs computed by ISTAT for the same products;
2) To compare the performances of the different index formulas discussed in the previous section when computed on scanner data (specifically, a comparison of the Lasperyes, Paasche, and GEKS indices).

We start in Figure 6.3 by comparing the monthly and weekly GEKS price indices computed on our dairy scanner data set with the dairy CPI computed by ISTAT using a chain-type fixed basket Laspeyres index.[7] This comparison

---

[6] For a more detailed explanation of the RWGEKS, please refer to Ivancic et al (2009).

[7] ISTAT computes monthly chain-type Laspeyres indices using as base period December of the previous year. The value of the index in the reference period is obtained by simple multiplication of

The Use of Scanner Data for Measuring Food Inflation

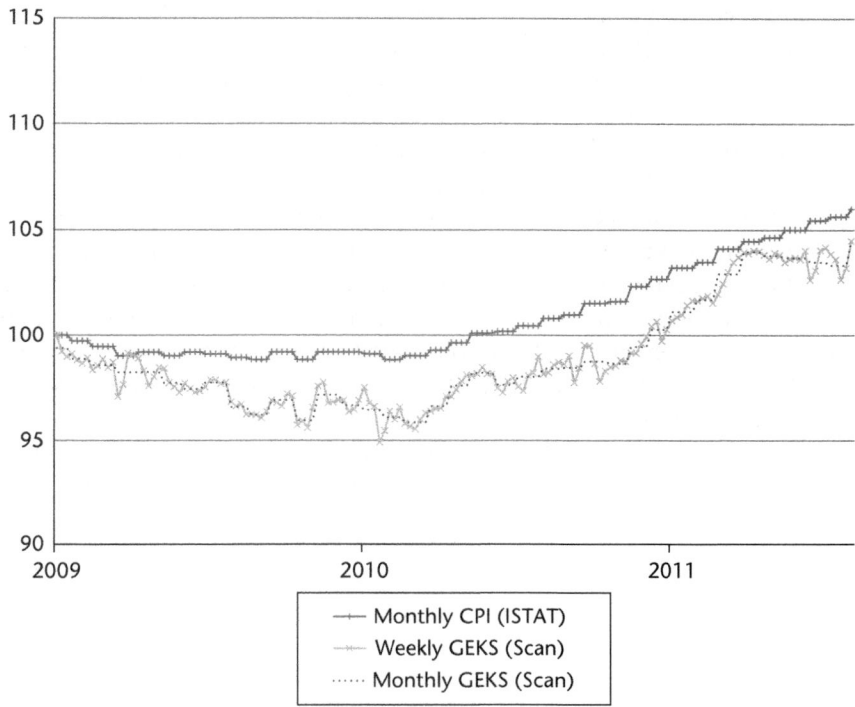

Figure 6.3. Comparison of dairy price indices computed on different data sets (2009–11)

Source: Own elaboration on Symphony IRI and ISTAT data

Note: The base period of the monthly indices is January 2009. The base period for the weekly index is the first week of January 2009.

must be taken with some caution, since the scanner data set does not include some important categories of products for the Italian dairy sector (mainly hard and semi-hard cheeses). Thus, the fact that the GEKS indices show stronger reductions in the first half of the period observed, and stronger increases in the second half, may be partly related to this difference in the composition of the product basket. However, it is clear that the GEKS index better captures the volatility of dairy prices, which is likely to be related to the stronger heterogeneity in the composition of the basket and to the pricing strategies of manufacturers and retailers, mainly promotion strategies.

The same type of comparison is carried out by product category, where the heterogeneity of the reference baskets is due just to the different number of items considered in the calculations, since the standard CPI procedure

---

the index of month *m* by the December indices computed for all the previous years, until the reference year.

considers changes in prices of a few reference items, while the indices computed on the scanner data set consider all items sold in a given period. In order to make the comparison as homogeneous as possible, we have computed the Laspeyres price index also on our scanner data set, considering only the year 2011, since the corresponding indices by ISTAT are available only for that year. Of course, the calculation of the Laspeyres index on scanner data requires some approximation in order to construct a proper fixed reference basket, since at the item level the basket of products changes very often, with many entries and exits. Thus, we aggregated groups of items (i.e., different packaging sizes for the same brand, different brands for the same manufacturer, etc.) in order to obtain a fixed basket of item groups that is sold over the whole observation period (one year). This makes the calculation of the Laspeyres index possible, using as weights the expenditure shares in the first week.

In Figure 6.4 we analyse the case of refrigerated milk. While the general trend of the three indices is similar and characterized by a rather smooth

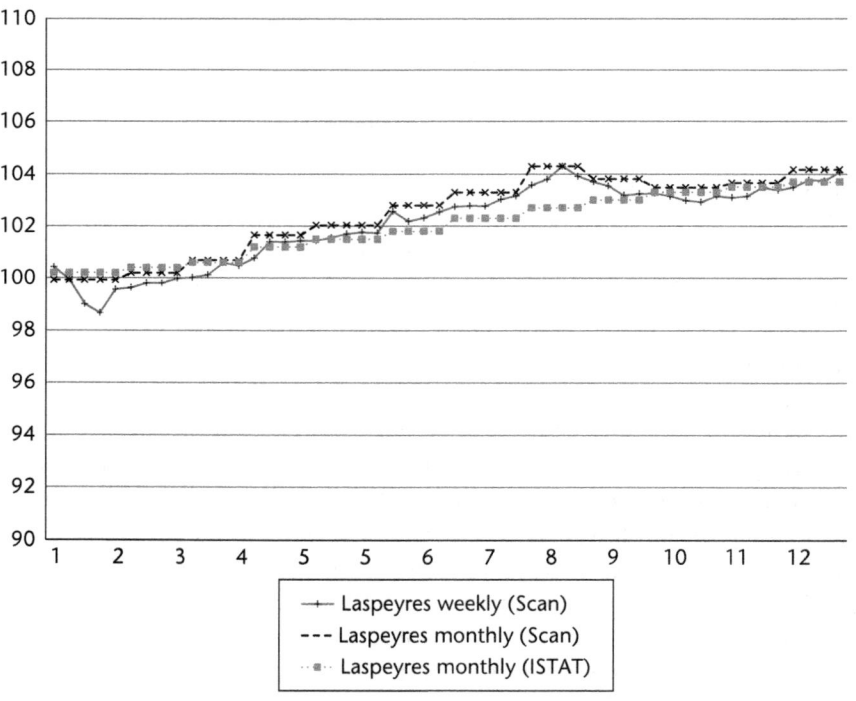

Figure 6.4. Comparison of Laspeyres refrigerated milk price indices computed on different data sets (2011)

Source: Own elaboration on Symphony IRI and ISTAT data

Note: The base period of the monthly indices is December 2010. The base period for the weekly index is the last week of December 2010.

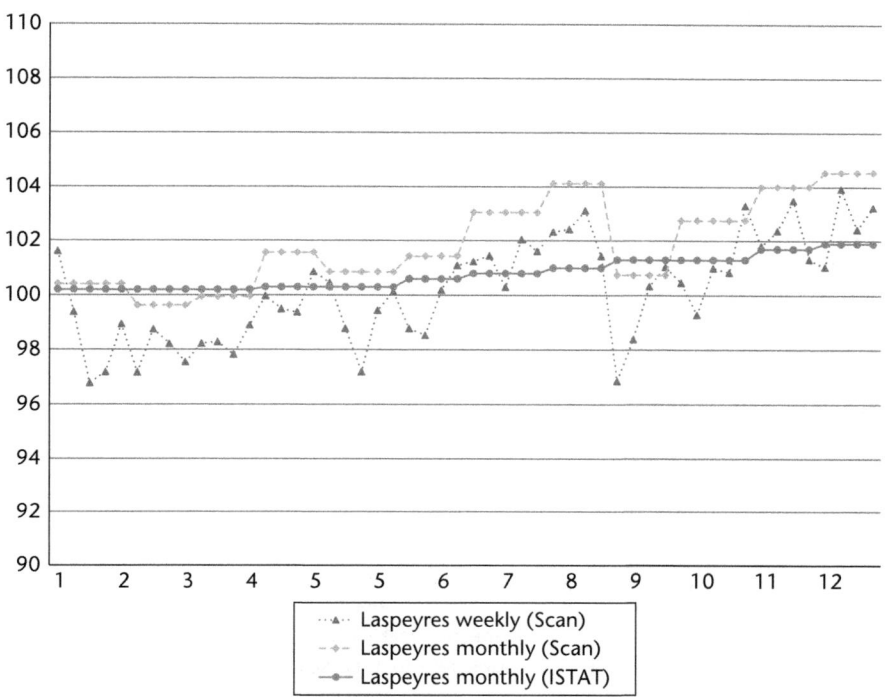

Figure 6.5. Comparison of Laspeyres UHT milk price indices computed on different data sets (2011)

Source: Own elaboration on Symphony IRI and ISTAT data

Note: The base period of the monthly indices is December 2010. The base period for the weekly index is the last week of December 2010.

increase, it is clear that with scanner data one can better capture the variability of prices over time, especially if weekly data are used. This general consideration is even clearer when we consider more volatile markets, such as UHT milk (Figure 6.5), butter (Figure 6.6), and yoghurt (Figure 6.7). In all three cases, the 'official' CPI displays a rather smooth increasing trend, while the Laspeyres indices computed on scanner data, both monthly and weekly, show a much stronger variability that is probably related to the aggressive promotion strategies carried out by manufacturers and retailers for these specific product categories.

For the same four products, we have also compared the performances of the traditional Laspeyres and Paasche formulas with the GEKS index, all computed on the same data and on a weekly basis for the whole three-year observation period. Again, the two fixed basket indices were computed over the item groups defined above, thus losing some variability, mainly due to

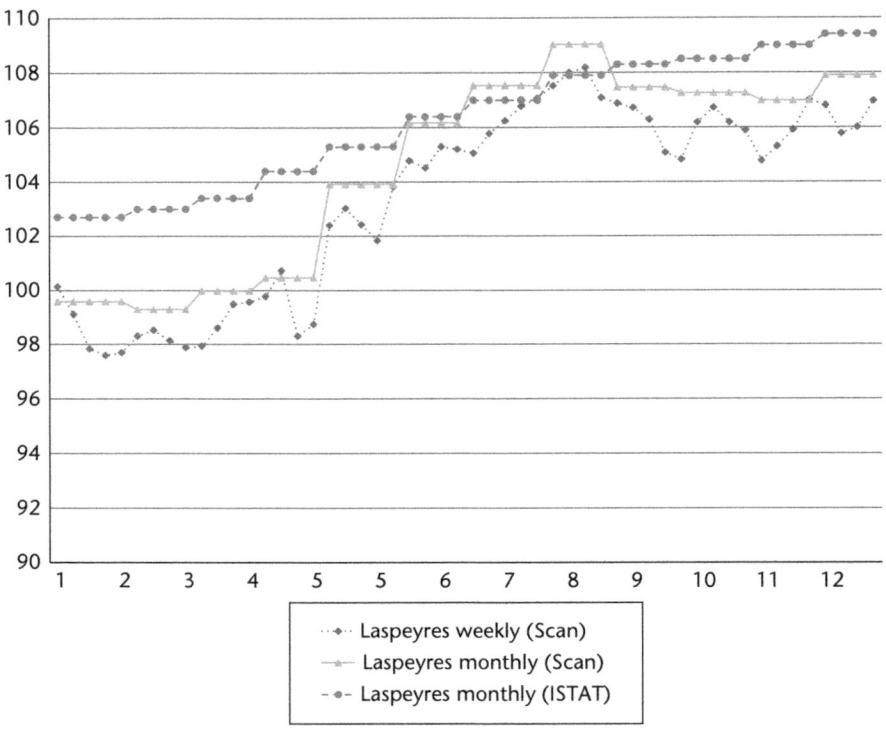

Figure 6.6. Comparison of Laspeyres butter price indices computed on different data sets (2011)

Source: Own elaboration on Symphony IRI and ISTAT data

Note: The base period of the monthly indices is December 2010. The base period for the weekly index is the last week of December 2010.

product entries and exits. If we analyse Figures 6.8 (refrigerated milk), 6.9 (UHT milk), 6.10 (butter), and 6.11 (yoghurt), we can see that volatility is adequately captured also by these indices, whose trends tend to be almost identical since the reference basket is the same. Only for yoghurt do we observe a relevant difference between the Laspeyres and the Paasche indices, which is likely to be related to the features of this market, which is extremely dynamic, with a strong competition among NBs and between NBs and PLs, such that in three years the shares of the item groups have changed quite drastically. In all four markets, the GEKS index shows a quite different trend, which is related both to the aggregation procedure described above, that is carried out only for the Laspeyres and Paasche indices, and to the possibility of properly accounting for the changes in the product basket over time. For refrigerated milk, the GEKS index stays constantly above the other two indices, while for the other

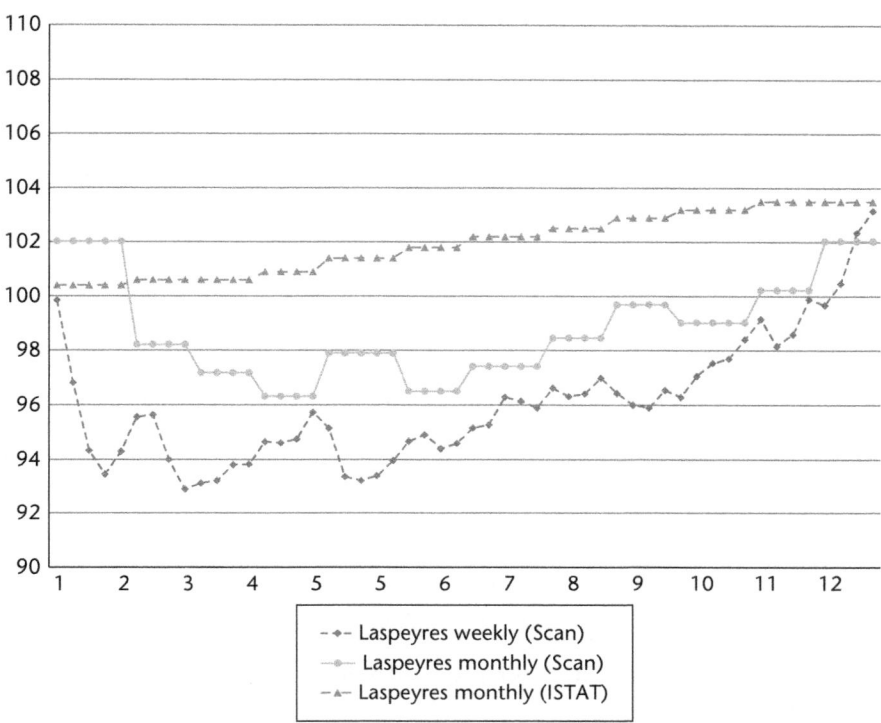

Figure 6.7: Comparison of Laspeyres yoghurt price indices computed on different data sets (2011)
Source: Own elaboration on Symphony IRI and ISTAT data
Note: The base period of the monthly indices is December 2010. The base period for the weekly index is the last week of December 2010.

products the trend is mixed (mainly below for UHT milk and butter, mainly above for yoghurt). In general, it is not easy to identify the causes of such differences. However, given the calculation procedure of the GEKS index, for which the products and the weights change in each single period, when the GEKS stays above it is probably because high-value products tend to gain market shares, while the opposite is true when the GEKS stays below.

Summarizing, this simple descriptive analysis of the behaviour of the indices leads us to the following conclusions:

- the indices computed on scanner data better capture the variability of prices over time, especially in highly volatile markets, such as many food markets;
- the indices computed on scanner data allow us to properly evaluate the impact of pricing strategies by manufacturers and retailers, especially aggressive promotion strategies;

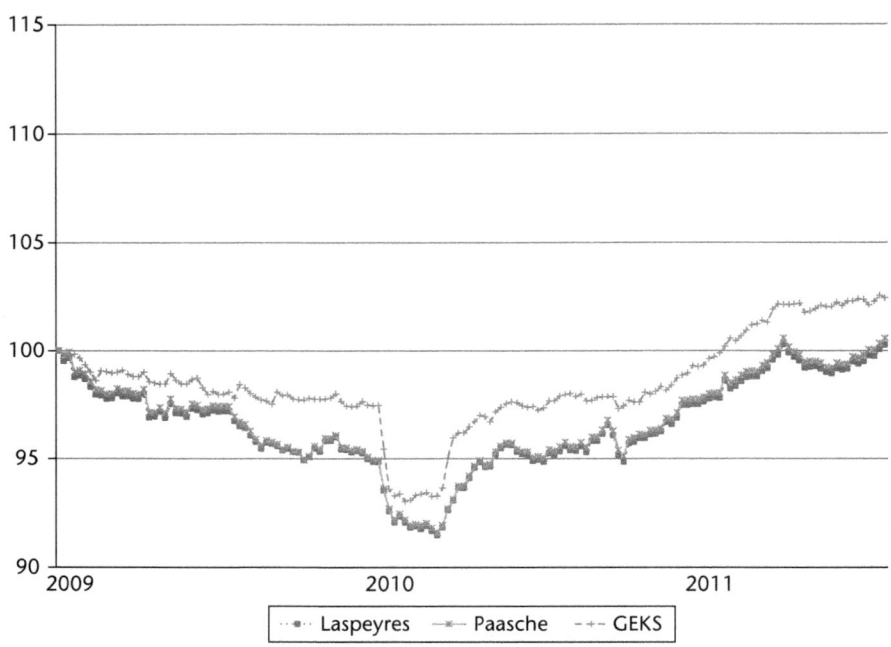

Figure 6.8. Comparison of different refrigerated milk price indices computed on scanner data (2009–11)

Source: Own elaboration on Symphony IRI data

Note: The base period of the monthly indices is January 2009. The base period for the weekly index is the first week of January 2009.

 – using superlative flexible basket indices like the GEKS, as compared to the traditional Lapseyres and Paasche indices, allows us to properly evaluate the impact of entry and exit of new products and of changing market shares, since this impact can be quite sizeable.

6.7 Conclusions

In recent decades the CPI's role in the measurement of price inflation has been changing. From a simple measure of the price change of a representative basket of goods, the CPI has been challenged as a measure of the true cost of living.

Five different sources of biases have been found to affect the 'traditional' CPI computation: substitution bias, outlet substitution bias, new product bias, quality change bias, and elementary index bias. Thus alternative indices, such

The Use of Scanner Data for Measuring Food Inflation

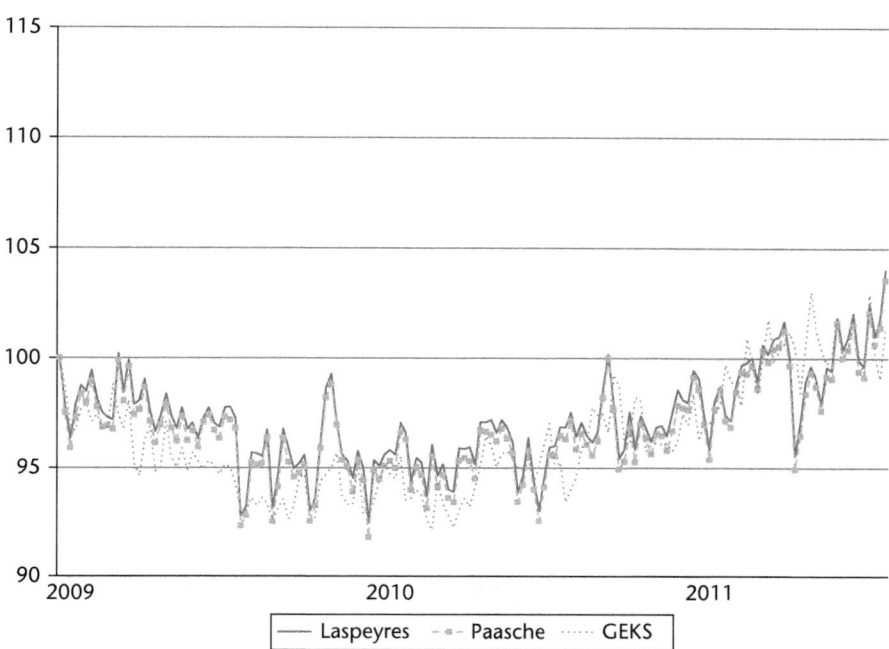

Figure 6.9. Comparison of different UHT milk price indices computed on scanner data (2009–11)

Source: Own elaboration on Symphony IRI data

Note: The base period of the monthly indices is January 2009. The base period for the weekly index is the first week of January 2009.

as superlative price indices, have been proposed. Furthermore, markets are highly heterogeneous with a large variety of goods and price variability, also due to promotion strategies and spatial competition among retailers, thus making the measurement of inflation more challenging.

High-frequency data, collected using scanner technology, can be good candidates for the CPI computation. They record prices and quantities for a very large number of goods, thus making it possible to account for substitution patterns and for the introduction of new products, and to select a more representative basket, given that they consider actual consumer behaviour. The advantages of using scanner data are sizeable, although they require a different methodological approach, since conventional price indices formulas applied to high-frequency data usually produce highly volatile indices and chain drift bias. Thus a viable solution is to resort to drift-free chain indices such as the Gini-Eltetö-Köves-Szulc (GEKS) index.

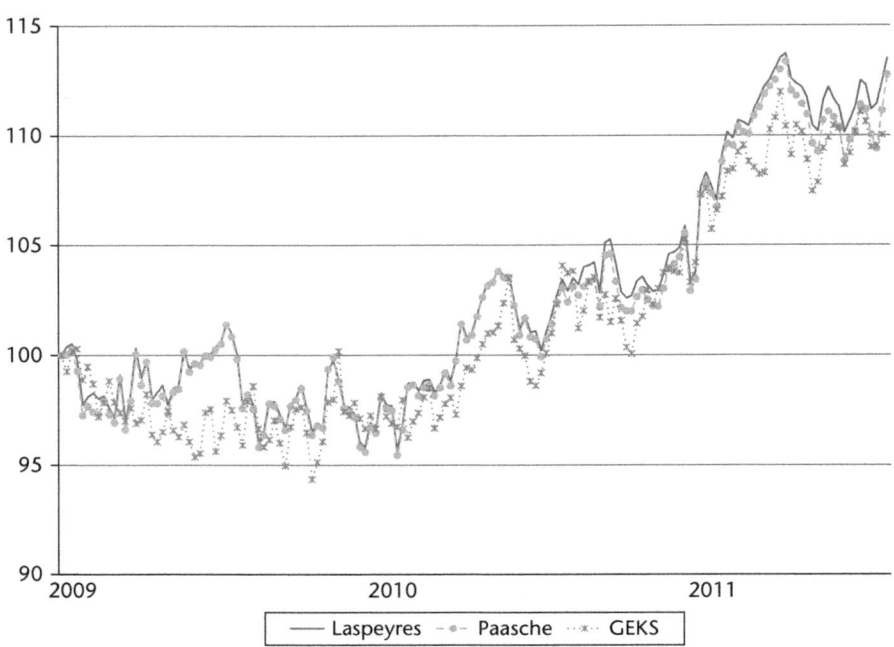

Figure 6.10. Comparison of different butter price indices computed on scanner data (2009–11)

Source: Own elaboration on Symphony IRI data

Note: The base period of the monthly indices is January 2009. The base period for the weekly index is the first week of January 2009.

We have used high-frequency scanner data for the Italian dairy market to compute price indices. A comparison of GEKS indices with the dairy CPI computed by the Italian Statistical Agency shows strong differences and indicates that, thanks to data heterogeneity and the inclusion of sales, the GEKS indices better capture volatility, especially when weekly indices are considered. This feature translates also to the Laspeyres and Paasche indices when computed on high-frequency data. However, the GEKS index, being a drift-free chain index, should be able to account only for the volatility linked to consumers' behaviour. Comparing the GEKS indices with the Laspeyres and Paasche indices indicates rather different price behaviour. From our analysis it turns out that a more precise picture of food inflation can be obtained by resorting to the GEKS index, which better accounts for pricing strategies by retailers and manufacturers, entry and exit of products, market dynamics, and variable market shares.

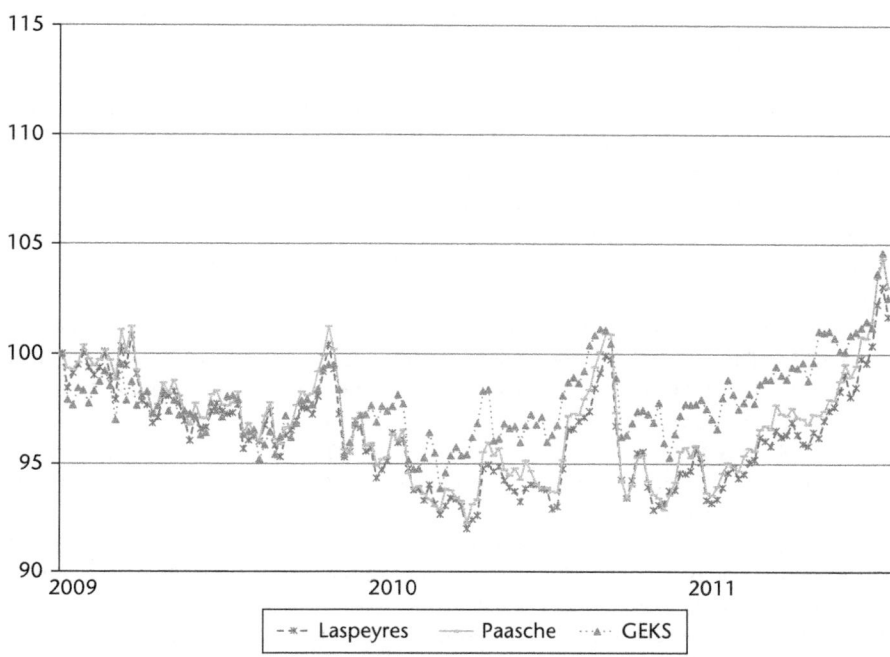

Figure 6.11. Comparison of different yoghurt price indices computed on scanner data (2009–11)

Source: Own elaboration on Symphony IRI data

Note: The base period of the monthly indices is January 2009. The base period for the weekly index is the first week of January 2009.

References

Abraham, K. G. (2003). Toward a Cost-of-Living Index: Progress and prospects. *Journal of Economic Perspectives* 17: 45–58.

Boskin, M. J., E. R. Delberger, R. J. Gordon, Z. Griliches, and D. W. Jorgenson (1996). Towards a More Accurate Measure of the Cost of Living. Final Report to the Senate, 4 December 1996.

Boskin, M. S., E. R. Delberger, R. J. Gordon, Z. Griliches, and D. W. Jorgenson (1998). Consumer prices in the Consumer Price Index and the Cost of Living. *Journal of Economic Perspectives* 12: 3–26.

Boskin, M. J. and D. W. Jorgenson (1997). Implication of Overstating Inflation for Indexing Government Programs and Understating Economic Progress. *American Economic Review* 87: 89–93.

Bradley, R., B. Cook, S. G. Leaver, and B. R. Moulton (1997). An Overview of Research on Potential Uses of Scanner Data in the U.S. CPI. Paper Presented at the International Conference on Price Indices, 1997, Voorburg, The Netherlands. Available at <http://www.ottawagroup.org>.

Braithwait, S. D. (1980). The Substitution Bias of the Laspeyres Price Index: An Analysis Using Estimated Cost-of-Living Indexes. *American Economic Review* 70: 64–77.

Dalen, J. (1997). Experiments with Swedish Scanner Data. Paper Presented at the International Conference on Price Indices, Voorburg, The Netherlands. Available at <http://www.ottawagroup.org>.

de Haan, J. and E. Opperdoes (1997). Estimation of the Coffee Price Index Using Scanner Data: the Choice of the Micro Index. Report (Department of Consumer Prices, Statistics Netherlands, Voorburg). Available at <http://www.ottawagroup.org>.

de Haan, J. and H. A. van der Grient (2011). Eliminating Chain Drift in Price Indexes Based on Scanner Data. *Journal of Econometrics* 161: 36–46.

Diewert, W. E. (1976). Exact and Superlative Index Numbers. *Journal of Econometrics* 4: 115–45.

Diewert, W. E. (1995). Axiomatic and Economic Approaches to Elementary Price Indexes. Discussion Paper No. 95-01, Department of Economics, University of British Columbia.

Diewert, W. E. (1998). Index Number Issues in the Consumer Price Index. *Journal of Economic Perspectives* 12: 47–58.

Feenstra, R. and M. Shapiro (2001, revised). High-Frequency Substitution and the Measurement of Price Indexes. Paper prepared for the Conference on Research in Income and Wealth on Scanner data and Price Indexes, 15–16 September 2000.

Ivancic L. (2005). Can Scanner Data Be Used to Reduce Substitution Bias in the CPI?: A Review of the Issues and Evidence. Center of Applied Economics Research, University of New South Wales. Working Paper.

Ivancic, L., W. E. Diewert, and K. J. Fox (2009). Scanner Data, Time Aggregation and the Construction of Price Indexes. Discussion Paper No. 09-09, Department of Economics, University of British Columbia.

Ivancic, L., W. E. Diewert, and K. J. Fox (2011). Scanner Data, Time Aggregation and the Construction of Price Indexes. *Journal of Econometrics* 161: 24–35.

Kliesen, K. L. (1997). Critiquing the Consumer Price Index. *The Regional Economist*. Available online at <http://www.stlouisfed.org/publications/re/articles/?id=1785>.

Nakamura, A. O., E. Nakamura, and L. I. Nakamura (2011). Price Dynamics, Retail Chain and Inflation Measurement. *Journal of Econometrics* 161: 47–55.

Reinsdorf, M. (1999). Using Scanner Data to Construct CPI Basic Component Indexes. *Journal of Business and Economic Statistics* 17: 152–60.

Richardson, D. H. (2000). Scanner Indexes for the CPI. Paper Presented at the Sixth Meeting of the International Working Group on Price Indices, 2–6 April 2001, Canberra, Australia.

Rodriguez, J. and F. Haraldsen (2006). The Use of Scanner Data in the Norwegian CPI: The 'New' Index for Food and Non-alcoholic Beverages. *Economic Survey* 4: 21–8.

Schut, C. (2001). Using Scanner Data to Compile Price Indices: Practical Problems. Paper presented at the Joint Statistical Commission and Economic Commission for Europe/International Labour Organisation (ECE/ILO) Meeting on Consumer Prices.

Silver, M. and B. Webb (2002). The Measurement of Inflation: Aggregation at the Basic Level. *Journal of Economic and Social Measurement* 28: 21–35.

Szulc (Schultz), B. J. (1983). Linking Price Index Numbers. In W. E. Diewert and C. Montmarquette (eds), *Price Level Measurement*. Statistics Canada, Ottawa, 537–66.

Triplett, J. E. (2003). Using Scanner Data in the Consumer Price Index: Some Neglected Conceptual Considerations. In R. C. Feenstra and M. D. Shapiro (eds), *Scanner Data and Price Indexes*. Chicago: University of Chicago Press, 151–62.

Walsh, C. M. (ed.) (1901). *The Measurement of General Exchange Value*. New York: Macmillan.

7
Price Transmission in Modern Agricultural Value Chains: Some Conceptual Issues

Johan Swinnen and Anneleen Vandeplas

7.1 Introduction

Recent changes in global food prices have had different impacts on domestic prices across countries, and on prices at different stages of the value chain within countries. This has resuscitated interest amongst policymakers in the issue of price transmission, and its implications for producer and consumer welfare.[1] Related to this, there has been significant debate on the extent to which high prices benefit producers, in particular in developing countries. Some have argued that consumers in developing countries have been hurt by increasing food prices while producers have been unable to benefit from them, increasing hunger and poverty in aggregate.[2] Price transmission issues have also stirred policy discussion in richer countries. Evidence shows that in the EU, on average, producer prices have varied more than consumer prices

---

[1] There is an intriguing difference in the policy reactions to high and low prices before and after 2007, as documented by Swinnen (2011) and Swinnen and Squicciarini (2012).

[2] Empirical evidence also shows a mixed picture. FAO (2009) argues that in African countries such as Kenya and Mozambique, consumer prices rose significantly, while farm-gate prices remained flat. A review of cereal markets in 52 countries over the period 2007–11 finds that transmission of price shocks from the world market to domestic markets varied from 50% to 100% (Sharma, 2011). Based on an analysis of price volatility of agricultural and food commodities in Africa, Minot (2012) finds that only 7 out of 17 prices have been more volatile since 2007, while 17 show significantly less volatility. Jacoby (2013) argues that the price hikes benefited poor rural households in India through positive wage effects for agricultural labour. Headey (2011) and Verpoorten et al. (2013) find that, on average, self-reported food security improved in net food producing countries and in rural areas over the same period.

(Swinnen et al., 2014). Moreover, it has been argued that when EU agricultural commodity prices were on the rise in 2007/8, these increases were passed on to consumers, but when prices declined again in 2008/9, these price declines were less than fully transmitted to consumers, hindering demand recovery and exacerbating the negative effect of declining producer prices on farm households (European Commission, 2009). The European Commission (2009) argued that the observed discrepancies between producer and consumer price developments reflected 'structural weaknesses in the system, such as the number of intermediaries operating along the chain and the competitive structure' and 'pervasive inequalities in bargaining power between contracting parties', and established a 'Task Force Food' within DG Competition to oversee competition in the food sector in 2012.

The transmission of price shocks has been studied extensively in the literature, and from different angles. Transmission of price shocks at the consumer level (e.g., triggered by a demand shock) to producers in domestic markets—and vice versa—is referred to as *vertical* price transmission. Transmission of price shocks in the world market to domestic markets—and vice versa—is referred to as *spatial* price transmission.

Imperfections in spatial price transmission have been attributed to different factors such as government intervention in markets (e.g., import tariffs and price stabilization measures), transport and marketing costs, the degree of processing, market structure, and consumer preferences (e.g., if imported products are imperfect substitutes for domestic products) (see, amongst others, Rapsomanikis, 2011).

Imperfect vertical price transmission, on the other hand, and in particular when it is asymmetric, has often been interpreted as providing evidence of the exercise of market power by processing companies and/or retailers, enabling them to capture supply chain rents and reduce social welfare (Wohlgenant, 2001; Meyer and von Cramon-Taubadel, 2004).[3]

However, this is not a consensus argument. There are a number of studies refuting the direct link between the degree of price transmission and market power, arguing that one should account for the incidence of vertical coordination in supply chains, the existence of increasing returns to scale, risk mitigating behaviour by intermediaries, and the degree of processing (McCorriston et al., 2001; Wohlgenant, 2001; Weldegebriel, 2004; Wang et al., 2006). For example, Wang et al. (2006) show that with increasing returns to scale, price transmission in the presence of market power can be

---

[3] The existing literature tends to focus on the effects for consumer welfare and generally assumes a positive correlation between the degree of downstream vertical price transmission and consumer welfare—as a lower degree of price transmission would attest to a greater share of the rents being captured by powerful intermediaries in the chain.

weaker than, identical to, or even stronger than in the competitive markets case. Meyer and von Cramon-Taubadel (2004) argue that apart from market power, menu costs, product storage and perishability characteristics, and price policies (e.g., price floors) may result in asymmetric price transmission as well. Bonnet and Villas Boas (2013) show that if consumers react less strongly to price increases than to decreases, price transmission will also be asymmetric. In an empirical analysis of the German dairy sector, Holm et al. (2012) find substantial asymmetric price transmission, but they do not find a correlation with the strength of a brand (a proxy for market power), and they find that the impact on profits is limited.[4]

Most of the studies of price transmission focus on the transmission of (global and domestic) producer prices to consumer prices (Chang and Griffith, 1998; von Cramon-Taubadel, 1998; Goodwin and Holt, 1999; Bonnet and Réquillart, 2012; Davidson et al., 2012; Holm et al., 2012). Less attention has been paid to studying upstream vertical price transmission, that is, the effects of price shocks originating in consumer markets on producer prices (Wohlgenant, 2001). These effects are important, however, as globalization and income growth have brought about important shifts in consumer demand, which are transforming supply chains all over the world, with important effects for local producers.

Moreover, a large share of the existing literature on price transmission is empirical in nature and builds upon theoretical work by McCorriston and Sheldon (1996) and McCorriston et al. (1998). Relatively little attention has been focused on refining the theoretical assumptions underlying these empirical analyses, to account for important institutional features such as the changing architecture of markets. This chapter attempts to fill this gap by theoretically examining how exogenous consumer price shocks (triggered, for instance, by income changes, global shocks, or by changes in consumer preferences) are transmitted to producer prices, taking into account the particular nature and institutional characteristics of ('modern') agricultural and food markets and supply chains.

Both Swinnen and Vandeplas (2010, 2011), who analyse global supply chains in emerging and developing countries, and Sexton (2012) and Crespi et al. (2012), who focus on US markets, emphasize that production for 'modern' markets requires consistency and strict adherence of products and production processes to quality and safety standards. This typically implies important investments by suppliers, but it often also entails substantial

---

[4] Swinnen and Vandeplas (2010) review existing studies on the welfare effects of concentration in supply chains, which show that potential negative welfare effects of concentration can be offset by efficiency gains due to scale economies, reduced transaction costs, enhanced incentives for R&D, countervailing bargaining power, and the sustainability of vertical coordination.

complementary investments for buyers through a vertically coordinated supply chain. Economic analysis needs to explicitly consider this 'new architecture of modern agricultural markets' as it has important implications for efficiency and equity. For example, Sexton (2012) argues that if high-quality supply chains require vertical coordination, and if buyer sunk costs and transaction costs in finding new suppliers are significant, monopsonistic or oligopsonistic buyers may pay suppliers as much as—or even more than—competitive markets. Similarly, Swinnen and Vandeplas (2011) show that even with unequal bargaining power, buyers may pay suppliers 'efficiency premia' to ensure quality supplies in environments with factor market imperfections and weak bargaining power. Crespi et al. (2012) show that if vertical coordination is pervasive, policymakers dealing with issues such as market power based on traditional models of agricultural markets (as highly competitive spot markets of homogeneous products) may devise policies running counter to their own objectives.

In this chapter we integrate these arguments to study price transmission. We extend the model by Swinnen and Vandeplas (2011), integrating insights from Sexton (2012) and accounting for quality requirements, contract-specific investments, factor market imperfections, and imperfect contract enforcement institutions. We use this model to show that price transmission depends on the nature of vertical coordination and different types of transaction costs in the supply chain. Our analysis shows that price transmission is mostly non-linear in modern value chains. We also conclude that, contrary to what is often assumed in empirical research, weaker price transmission from consumer to producer prices does not necessarily imply a welfare loss for suppliers.

Our framework provides only a partial analysis, as it abstracts away from several issues that have been pinpointed by other authors as being important for price transmission (e.g., consumer demand, economies of scale in production, menu costs). As such, it is only a first step in the broader exercise of setting up a comprehensive theoretical framework that incorporates important features of modern markets and from which empirically testable hypotheses can be derived. At the same time, our model and the insights it offers can help to further nuance earlier findings and concerns about imperfect price transmission. It shows that one must go beyond merely analysing price statistics, by exploring the institutional set-up of value chains, before drawing any conclusions on possible welfare implications for players at different stages of the value chain.

This chapter is structured as follows. Section 7.2 describes the types and characteristics of contracting costs in modern, vertically coordinated value chains. Section 7.3 presents a conceptual framework that incorporates these features for the analysis of price transmission. Finally, Section 7.4 concludes.

7.2 Contracting Costs in Modern Value Chains

As explained in the introduction, production for 'modern' markets requires consistency and strict adherence of products and production processes to quality and safety standards. While there are important similarities between models focusing on rich countries and those analysing global emerging and developing markets, there are also differences between the two strands of literature, in particular regarding the nature of the transaction costs and associated market governance. Sexton and colleagues focus on the role of search and switching costs associated with quality and consistency requirements in developed economies. While these types of transaction costs are also increasingly important in emerging and developing economies, the latter are further characterized by major imperfections in rural factor markets and poor contract enforcement institutions. These give rise to additional transaction costs and different vertical coordination strategies. Adherence to quality and safety standards typically implies important investments by suppliers, but it often also entails substantial complementary costs for buyers.[5]

We focus on four types of contracting costs:[6] search (or switching) costs incurred in the process of identifying suitable producers; the costs of training suppliers to produce high-quality commodities, which may imply the use of new technologies or compliance with new standards; monitoring costs to ensure that suppliers apply the technology as has been recommended; and the costs of providing external inputs (e.g., fertilizers, credit, seeds, and/or other types of technology). We focus on these because (a) they have been described as important in the literature; (b) they differ in two important characteristics, which allows for the development of a classification of contracting costs along two dimensions; and (c) because these dimensions importantly influence the effects of the contracting costs on efficiency and equity of contracts. The two key dimensions are whether the nature of contracting costs implies certain cost advantages of working with a repeat supplier vis-à-vis working with a potential new supplier; and whether the supplier-specific investment has a value for the supplier outside of the existing contract.

7.2.1 Cost Advantage of Repeat Suppliers

Some contracting costs are the same for first-time and repeat suppliers alike since they need to be incurred for each production cycle. This is the case, for

---

[5] Like Sexton (2012) and Swinnen and Vandeplas (2011), we do not consider sunk costs in processing infrastructure, but focus on supplier-specific contracting costs.

[6] In our analysis, we use the concepts of (supplier-specific) 'buyer investment' and 'contracting costs' interchangeably, since we interpret (and model) them as the cost of the investment which the buyer needs to make in order to make a transaction possible.

instance, when buyers pre-finance external inputs for their suppliers. In many developing and emerging countries, processors provide their suppliers under contract with seeds, pesticides, and fertilizer in crop production and feed in livestock production (Gow et al., 2000; Birthal et al., 2005; Bellemare, 2012).

Another example is when the buyer has a system in place to monitor contract compliance—for example, in the form of a set of field officers who conduct field visits on a regular basis. By means of illustration, consider the following quote from Minten et al. (2009: 1733) on monitoring in high-value vegetable production in Madagascar:

> To monitor the correct implementation of the supplier contracts, the [processor] has...around 300 extension agents who are permanently on the payroll...Every extension agent, the chef de culture, is responsible for about 30 farmers. To supervise these, (s)he coordinates five or six extension assistants...that live in the village itself....During the cultivation period of the vegetables under contract, the contractor is visited on average more than once...a week...to ensure correct production management as well as to avoid 'side-selling.'...99% of the farmers say that the firm knows the exact location of the plot; 92% of the farmers say that the firm will even know...the number of plants that are on the plot. For some crucial aspects of the vegetable production process, representatives of the company will even intervene in the production management to ensure it is rightly done.

Other contracting costs are non-recurring, and only need to be paid at the start of a collaboration between a buyer and a supplier. Working with a repeat supplier, for which these costs have already been borne, is then cheaper than contracting a new supplier. An example is search costs, which are incurred to identify a suitable supplier. For example, Sexton (2012: 215) points out that 'transaction costs of engaging with repeat suppliers will likely be considerably less than transaction costs of locating and contracting with new suppliers'.

Another example is training costs. A new supplier typically needs to be trained to become familiar with buyer preferences and standards, and possibly with the use of new technologies. Being trained confers a cost advantage to repeat suppliers compared to potential new suppliers.

7.2.2 Value of Investment outside of the Contract

Another dimension of these contracting costs is their residual value to the contracted supplier outside the contract.[7] Some types of buyer investments do

---

[7] The costs under study are all 'supplier-specific'. Hence, once incurred, none of these costs have any residual value to the buyer/investor outside of the current contract relationship. This would have been different had we considered sunk costs in infrastructure by the buyer, for example.

not have a value for the supplier beyond the existing contract. Examples include search costs and monitoring costs.

Other types of costs may have an important residual value outside of the existing contract. For example, before being applied to crops, external inputs can be used by the supplier for other purposes (e.g., fertilizer use for other, non-contracted, crops) or sold on the secondary market. Even after being applied to crops, external inputs convey additional value to suppliers, which may be realized outside of the contract, for instance when the contracted supplier side-sells his produce to alternative buyers. Also training costs can have a value outside of the contract, depending on the degree of specificity of the training (Becker, 1962). Training increases a supplier's human capital. This is an intangible asset which can be used in other activities and which may have a long-lasting positive impact on the supplier's opportunity cost of labour.

7.2.3 Classification of Contracting Costs

Considering the two dimensions discussed above, we can classify contracting costs as four 'types'—as illustrated in Table 7.1. Monitoring costs and the costs of providing external inputs are recurring costs and do not provide a cost advantage to repeat suppliers; search costs and training costs are non-recurring and therefore do provide a cost advantage to repeat suppliers. External inputs and training have a value to the supplier outside the contract, while monitoring and search costs do not.

For didactic purposes, we consider 'pure' forms of the four types of contracting costs in the rest of the paper. In reality, of course, the boundaries between these four types are not always clear-cut. Some of the contracting costs may have mixed characteristics. For example, training may be a combination of an initial investment at the start of a collaboration between a buyer and a supplier, supplemented by regular training sessions, possibly at the beginning of each production cycle, in which suppliers receive updates of best practices. In such cases, the relative cost advantage of a repeat supplier to a new supplier depends on the relative importance of initial training sessions vis-à-vis the updates. Similarly, Dries et al. (2009) document how some of the investments

Table 7.1. A typology of contracting costs

		Value outside of the contract (for supplier)	
		$\alpha = 0$	$\alpha > 0$
Cost advantage of repeat supplier (for buyer)	$\gamma = 0$	Monitoring costs	External inputs
	$\gamma > 0$	Search costs	Training costs

in external inputs in the Eastern European dairy sector can be considered as non-recurring investments (e.g., cheap loans for equipment and herd upgrading), while other investments, such as feed concentrate, are incurred for each production cycle. In this chapter, when we talk about 'external inputs' we have the latter in mind, that is, those costs that are incurred for each production cycle.

7.3 A Model of Modern Value Chains

Consider the case where a 'supplier' (for example, a farm) can sell products to a 'buyer' (for example, a trader or a retailing or processing company).[8] Assume the buyer can offer a contract to the supplier, which includes the conditions (time, amount, and price) for purchasing the supplier's products, and selling these products (possibly after processing) to consumers—either domestically or internationally—in a quality-differentiated product market at a unit price p_h. We assume that consumer demand is perfectly elastic at price p_h and that this implies a quality premium for the buyer over his variable production costs.[9] This quality premium may, for instance, result if the buyer is a 'gatekeeper' to the high-value market (Inderst and Mazzarotto, 2008);[10] or if consumers are paying a 'quality-assuring price' (Klein and Leffler, 1981).

To produce q units of a high-quality product, the supplier needs to invest an amount l of own resources (e.g., labour or land). We assume the supplier's opportunity cost of these own resources is $\bar{l}$. If, for example, his best alternative use of these resources were to produce 1 unit of another product, then, $\bar{l} = p_l$ with p_l the price of the alternative product—which could be a low-quality product for the local spot market.

The production of high-quality agricultural products requires the buyer to invest an amount of contracting costs k, with $\bar{k}$ the buyer's opportunity cost of this investment. Two parameters, γ and α, capture the two dimensions of contracting costs defined in Section 7.2. The parameter γ represents the cost advantage for the buyer of working with repeat suppliers vis-à-vis new suppliers. The effective investment costs are $(1-\gamma)\bar{k}$ where $\gamma = 0$ for new suppliers or for recurring investments, which need to be done for each production cycle, irrespective of the supplier; and $\gamma = 1$ for repeat suppliers, when investments are entirely consisting of non-recurring costs. Hence, $\gamma \in [0;1]$.

---

[8] Our basic set-up is based on Swinnen and Vandeplas (2011).
[9] Or, as in Klein and Leffler (1981), over his 'salvageable' production costs—whether fixed or variable.
[10] Take, for example, the case of fruits and vegetables exporters in sub-Saharan Africa, where high food standard requirements increase the costs of entry for potential exporters, eventually leading to consolidation of the export supply base (Maertens and Swinnen, 2008).

Price Transmission in Modern Agricultural Value Chains

The residual value of the investment for the supplier outside the contract is defined as a fraction α of $\bar{k}$, its value within the contract. The less specific these investments are, the higher their value outside of the contract will be, that is, the higher α. If an investment does not have any value outside the contract for the supplier, $\alpha = 0$.

Based on these assumptions, we will now explore the conditions for contracting and the implications for price formation in a Nash bargaining framework. We start off by considering the case of perfect contract enforcement, where contracts are enforced at no additional cost. Next, we look into the case where there is no external contract enforcement, and contracts need to rely on internal enforcement. In other words, they need to be made self-enforcing.

7.3.1 Price Transmission with Perfect Contract Enforcement

If formal contract enforcement institutions work well and are costless, the only condition for contracts to be realized is that both agents' participation constraints are satisfied. For the supplier, this means that his income Y must cover the opportunity cost of his own resources:

$$Y \geq \bar{l}. \tag{1}$$

For the buyer, it means that his income Π must cover at least his opportunity cost of the capital invested in the contract:

$$\Pi \geq (1-\gamma)\bar{k}. \tag{2}$$

The contract surplus S is the net value created by the contract after subtraction of the opportunity costs of all invested resources:

$$S = p_h q - (1-\gamma)\bar{k} - \bar{l}. \tag{3}$$

In a Nash bargaining framework, this surplus will be shared between buyer and supplier according to a fixed sharing rule β with $0 \leq \beta \leq 1$ the share accruing to the supplier.[11] Both participants to the contract receive the opportunity cost of the resources they invest in the contract, topped up with a share of the contract surplus determined by β. The contract (Y, Π) then implies the following pay-offs:

---

[11] The determination of β is a question which has received a lot of attention in the theoretical literature but has not yet been fully resolved (see, e.g., Doyle and Inderst, 2007; Swinnen and Vandeplas, 2011). One part of the literature argues that β should be 0.5 in the case of perfect information and in the absence of uncertainty (e.g., Nash, 1953); while another part of the literature argues that in the real world β may reduce to 0 in a context of extremely unequal bargaining positions (e.g., Svejnar, 1986).

$$\begin{cases} Y^0 = \bar{l} + \beta(p_h q - (1-\gamma)\bar{k} - \bar{l}) \\ \Pi^0 = p_h q - Y \end{cases} \quad (4)$$

where Y^0 is the supplier's pay-off under perfect enforcement, and Π^0 the buyer's pay-off. The resulting producer price p^0 is easily obtained by dividing Y by q:

$$p^0 = [l + \beta(p_h q - (1-\gamma)\bar{k} - \bar{l})]/q \quad (5)$$

If we assume that the supplier's fallback option is to produce low-quality goods ($\bar{l} = p_l$), and that there is no output effect from shifting from low-quality production to high-quality production ($q = 1$), then Equation (5) reduces to:

$$p^0 = p_l + \beta[(p_h - p_l) - (1-\gamma)\bar{k}] \quad (6)$$

which is equivalent to $p^0 = p_l + \beta.S$ if $S > 0$. Note that if $S \leq 0$, there are no gains to be made from high-quality production, and the supplier will produce for the low-quality market, with $p^0 = p_l$.

The second part of the right-hand side term in Equation (6) shows how the producer price is a function of the difference in the price of low-value and high-value products $(p_h - p_l)$, contracting costs $(1-\gamma)\bar{k}$, and the sharing rule β. The producer price is positively (and linearly) related to the consumer price and to the sharing rule. The higher the consumer price for high-quality products (p_h) and the higher β, the higher p^0.

We now look at how shocks in consumer prices are transmitted to prices at the supplier level. Price transmission can be defined as $\tau = \partial p/\partial p_h$ and is illustrated in Figures 7.1(a) and 7.2(a), with $p^0 = p_l$ in price region A where $p_h \leq p_l + (1-\gamma)\bar{k}$ (meaning $S \leq 0$). In this price region, price transmission is zero. By producing low-quality goods, the supplier is shielded against shocks in the high-quality market (although they may be prone to shocks in the low-quality market).

If $p_h > p_l + (1-\gamma)\bar{k}$, $S > 0$, high-quality contracting takes place. As the producer price is determined through a bargaining process, an exogenous price shock Δp_h in the consumer price for high-quality goods will be transmitted to the producer only partially. The degree of price transmission equals the share of the surplus that the producer gets: $\tau = \partial p^0/\partial p_h = \beta$.[12]

---

[12] Price transmission or pass-through can also be defined differently, namely as the elasticity of p^0 to p_h, hence as $\frac{\partial p^0}{p^0} * \left(\frac{\partial p_h}{p_h}\right)^{-1}$. If $\tau = \partial p^0/\partial p_h = \beta$, this elasticity will be equal to $\beta * \frac{p_h}{p^0}$. Its magnitude will depend on the relative size of p_h vis-à-vis p^0: with larger p_h, smaller p_l, and larger contracting costs $((1-\gamma)\bar{k})$, $\frac{p_h}{p^0}$ is larger and hence pass-through as well. If p_l makes up a major

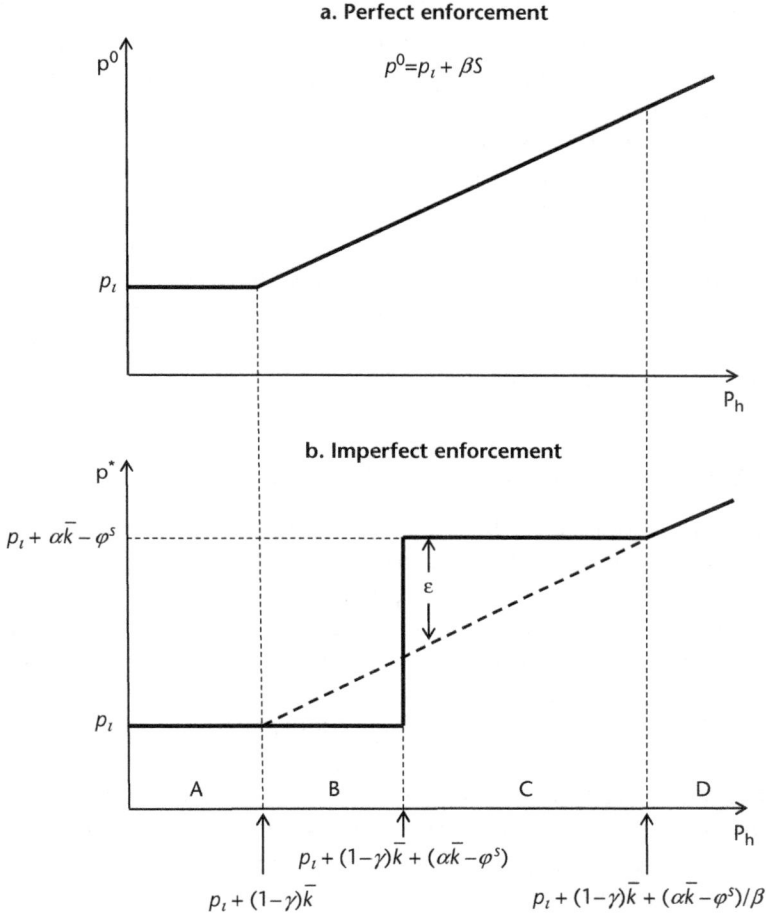

Figure 7.1. Relationship between producer and consumer prices

This, of course, is not surprising since a change in the consumer price affects the surplus and the change in the surplus is distributed in the same way as the initial surplus. With perfect enforcement, price transmission is not affected by contracting costs.

The process of bargaining for price determination partially isolates the farmer from the market. The lower β, the more strongly he is shielded from price shocks—both when prices go up and when they go down. It reduces the negative effect when prices fall, but it also reduces his gains when prices rise.

component of p^o, an increase in p_h may not have a major impact on p^o. This does not mean, however, that the producer is worse off.

In the extreme case where the surplus share of the supplier is zero ($\beta = 0$), the supplier is pushed back to his opportunity cost, and the buyer is the residual claimant of the full surplus. In this case, there will be no price transmission at all to the supplier level. Every change in the surplus (either an increase or a decrease) is absorbed by the buyer.

In summary, with perfect contract enforcement, the larger the surplus share β which is appropriated by the supplier, the stronger the degree of price transmission. Hence, within this scenario the traditional logic holds, that the stronger the degree of price transmission is, the stronger supplier benefits are: there is a positive correlation between β and p^0 (or Y^0). Note, however, that this also implies that the suppliers are more exposed to price volatility and that they will lose more when prices decline.

7.3.2 Price Transmission with Imperfect Contract Enforcement

If contracts are not perfectly enforceable, the outcome may be different. Opportunistic behaviour may lead to hold-ups if one of the agents has an attractive alternative to contract compliance. In particular, if the supplier can use the investment costs borne by the buyer to realize more value outside the contract than within the contract, and if contract enforcement is imperfect, a supplier will be tempted to violate the contract.

In practice, contract breach by the supplier can take many forms. For example, in the case where a buyer pre-finances the supplier's input costs, the latter can divert these received inputs to other uses, such as selling them or applying them to other production activities. An alternative way to hold up the buyer is when the supplier applies the inputs to the crops, as agreed in the contract, but then sells the high-quality output to an alternative buyer, who may or may not value the product as much as the contracted buyer.

In the case of training costs, opportunistic behaviour can arise in a similar way. Instead of applying his own resources (land and labour) in combination with the received training to produce high-quality goods, the supplier can use his training to earn additional income, for example by using his time and new skills in different production activities, or by producing high-quality goods for other buyers.

We assume that in case of contract breach, the supplier can realize his opportunity cost of labour, in addition to a fraction α of $\bar{k}$, with $\alpha\bar{k}$ the value of the buyer's investment which the supplier can realize outside the contract. The higher α, the more attractive contract breach will be. We also assume that by doing this the supplier will incur a cost φ^s. This cost can be interpreted in several ways: it can reflect a reputational cost, the loss of social capital, or the loss of future business opportunities (see, e.g., Keefer and

Knack, 2005 for a review).[13] The supplier payoff in case of contract breach is thus $\bar{l} + a\bar{k} - \varphi^s$.

Consider the extreme case that there is no external enforcement of contracts. In this case, contracts have to rely on internal enforcement: they need to be made self-enforcing. A self-enforcing contract requires that the supplier's contract income Y must cover at least his potential income from non-compliance with the contract. This condition constitutes the supplier's incentive compatibility constraint. Hence, in addition to the supplier's and the buyer's respective participation constraints (see Conditions (1) and (2)), the contract must satisfy the supplier's incentive compatibility constraint:

$$Y \geq \bar{l} + a\bar{k} - \varphi^s. \tag{7}$$

A self-enforcing contract (Y, Π) then implies the following incomes:

$$\begin{cases} Y = max[\bar{l} + \beta(p_h\, q - (1-\gamma)\bar{k} - \bar{l}); \bar{l} + a\bar{k} - \varphi^s] \\ \Pi = p_h q - Y, \end{cases} \tag{8}$$

with Y the supplier's income and Π the buyer's income from the transaction. The supplier price, p^*, in this contract is:

$$p^* = \frac{1}{q} max[\bar{l} + \beta(p_h\, q - (1-\gamma)\bar{k} - \bar{l}); \bar{l} + a\bar{k} - \varphi^s]. \tag{9}$$

The first term in the maximand in Equation (9) is the supplier price under perfect enforcement (see Equation (6)). This constitutes the lower bound to the supplier price under imperfect enforcement. The supplier price may be higher, however, if the supplier has an attractive option outside of the contract, once the buyer has made the required investment. In particular, if α is sufficiently high (such that $a\bar{k} - \varphi^s > \beta\,(p_h\,q - (1-\gamma)\,\bar{k} - \bar{l})$), the second term of the maximand in Equation (9) will bind and the producer price under imperfect enforcement will exceed the producer price under perfect enforcement.

If we assume the supplier's best alternative option is to produce one unit of a low-quality product ($\bar{l} = p_l$), and if $q = 1$, Equation (9) can be rewritten as:

$$p^* = max[p_l + \beta((p_h - p_l) - (1-\gamma)\bar{k}); p_l + a\bar{k} - \varphi^s]. \tag{10}$$

---

[13] Swinnen and Vandeplas (2011) model an additional alternative for contract breach: if the supplier uses the acquired investment to produce a high-quality product, but sells it on better terms to an alternative buyer, one can show that a contract needs to fulfil an additional condition to be self-enforcing, which depends on q and on the price a supplier can fetch on the spot market for the high-value product. This issue is ignored in this chapter for reasons of simplicity, but taking it on board should not affect our main conclusions.

Hence, the producer price under imperfect enforcement will be at least as high as under perfect enforcement, conditional of course upon the contract being sustainable. As for sustainability, the contract specified in Equation (8) should satisfy the buyer's participation constraint, which is $\Pi \geq (1-\gamma)\bar{k}$ (see Condition (2)). In combination with Conditions (1) and (7), this condition imposes a lower bound on p_h. Only if p_h is sufficiently high, is it possible to set the contract terms such that both agents' participation constraints as well as the supplier's incentive compatibility constraint are simultaneously satisfied. The specific conditions for contract feasibility are summarized in the following restriction on p_h:[14]

$$p_h \geq p_h^{min} = max\left(p_l + (1-\gamma)\bar{k},\ p_l + (1-\gamma+\alpha)\bar{k} - \varphi^s\right). \quad (11)$$

This condition captures two major reasons for potential contract failure. First, if $p_h q < p_l + (1-\gamma)\bar{k}$, the net surplus of the transaction will be negative, and there is no incentive for contract formation. Second, and more importantly, if $p_h \geq p_l + (1-\gamma)\bar{k}$ but smaller than $p_l + (1-\gamma+\alpha)\bar{k} - \varphi^s (= p_l + (1-\gamma)\bar{k} + (\alpha\bar{k} - \varphi^s)$ with $\alpha\bar{k} - \varphi^s$ the net benefits of contract breach), the contract surplus is positive, but the surplus is too small to allow the buyer to offer a price to the supplier which makes him comply with the contract. Under these conditions, the contract will not be realized, despite its potential positive contribution to social welfare. These conditions are represented by price regions A and B in Figure 7.1(b). When the potential surplus is negative (region A) or when the potential surplus is too low for the buyer to pay a sufficiently high price to the supplier (region B), there will be no contract and the income for the supplier will be his reservation income p_l.

Once consumer prices are high enough such that $p_h \geq p_l + (1-\gamma+\alpha)\bar{k} - \varphi^s$, contracting will occur and high-quality products will be produced. With imperfect contract enforcement, a buyer will have to pay his supplier a premium on top of the perfect enforcement outcome to prevent violation of the contract after the buyer has paid the contracting costs.

We refer to this premium as an 'efficiency premium' ϵ, which equals the difference between the supplier's price under (costless) perfect enforcement (p^0) and his price under costly enforcement (p^*): $\epsilon = p^* - p^0$. Making the contract 'self-enforcing' by paying an efficiency premium is a rational strategy for a buyer if it earns him a better pay-off than his outcome when being held up or his outcome when not engaging in a transaction with the concerned supplier, or any other one.

---

[14] We implicitly assume that the buyer can commit to the contract; see Swinnen and Vandeplas (2011) for when this is not the case.

Price Transmission in Modern Agricultural Value Chains

In Figure 7.1(b), this efficiency premium ϵ is represented by the difference between the full line and the dashed line. Notice that over price region C the producer price will be fixed at $p^* = p_l + \alpha\bar{k} - \varphi^s$. This implies that the efficiency premium will adjust to reflect the difference between p^* and p^0. This also means that price transmission is zero in this region (see Figure 7.2(b)). Note, however, that in region C producer prices are higher than they would be under perfect enforcement (and stronger price transmission), represented by the dashed line in Figure 7.1(b). In region B, however, the producer price with perfect enforcement is higher than with imperfect enforcement. Note that all this implies that there is no direct relationship between price transmission and producer incomes.

Once consumer prices increase further to where $p_h \geq p_l + (1-\gamma)\bar{k} + (\alpha\bar{k} + \varphi^f)/\beta$ (region D), producer prices will follow the increase in producer prices. Producer prices are $p = p_l + \beta\left(p_h - p_l - (1-\gamma)\bar{k}\right) = p^0$, the price with perfect enforcement. In this case price transmission is also β.

Figure 7.2 illustrates the variation in price transmission over the consumer price region. Price transmission (τ) is zero for price changes within regions A, B, and C. However, note that if the consumer price changes between regions B and C, there is a large, discontinuous price effect for producers. Similarly, if the consumer price shifts between regions C and D, there is a discontinuous effect.

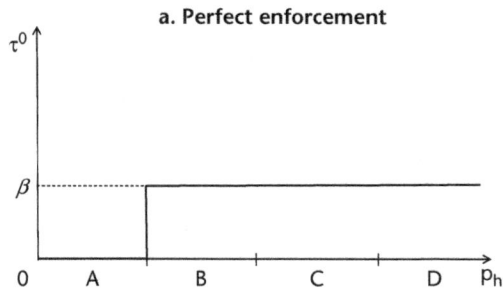

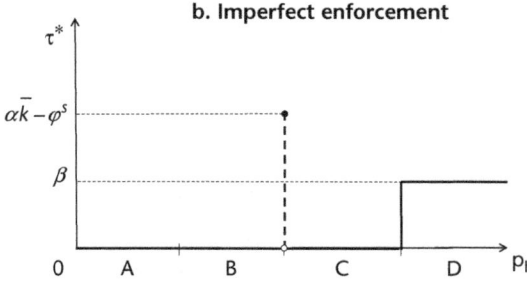

Figure 7.2. Price transmission (τ)

Finally, note that our discussion of the impact of consumer price shocks on suppliers involved in vertical contracts has focused on price shocks originating in the high-quality market. Welfare of suppliers involved in high-quality supply chains may also be affected by price shocks originating in the low-quality market. A price change in the low-quality market will affect p_l and it is obvious from Figures 7.1(a) and 7.1(b) that this would affect the shape of the high-quality price function.

7.3.3 The Impact of Contracting Costs on Price Transmission

The amount and nature of the contracting costs (reflected in the γ and α parameters) affect both the shape of the producer price function and the size of the different price regions, which together determine the process of price transmission. The impact of the different costs is illustrated explicitly in Figure 7.3. Panel 3(a) illustrates the impact of differences in α. Recall that α is an indicator for the value a supplier can realize based on the buyer's contract-

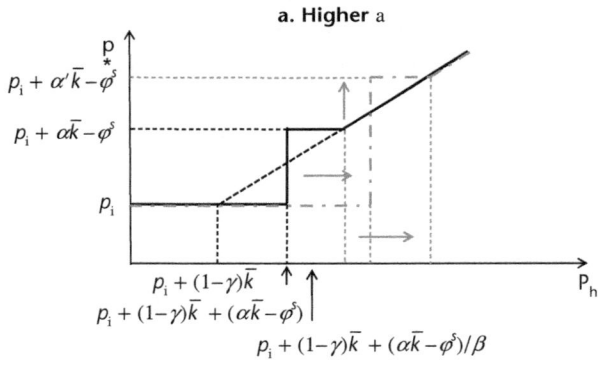

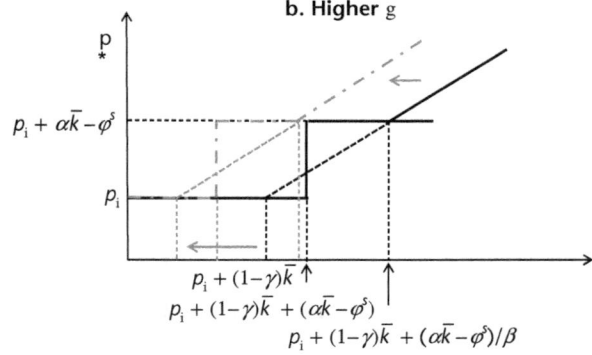

Figure 7.3. Impact of contract costs

specific investment outside of their joint contract. A higher α implies more benefits for a supplier from the associated (higher) efficiency premium. This is reflected in the upward shift of the function for the C region, and in the enlargement of the C region, for which the price transmission is zero. However at the same time a higher α makes internal contract enforcement harder. This is reflected in the rightward shift of the function and the enlargement of the B region where contracting is not possible.

Panel 3(b) illustrates the impact of γ, which captures the impact of search and training costs. With lower costs (and hence a higher γ), the price function shifts to the left. With lower costs there is more surplus, which makes contracting easier. This causes the shift to the left with a smaller A region. With more surplus in the contract, the first term in Equation (10) is larger and more likely to bind, reflected in a larger D region.

Note, however, that these changes in the function do not change the key findings that price transmission is discontinuous and inconsistent with the traditional logic, in which a positive correlation is assumed between producer welfare and the degree of price transmission.

7.4 Concluding Remarks

The empirical literature on price transmission usually assumes that perfect competition amongst buyers produces the best results for farmers and that perfect competition will result in perfect transmission of price shocks along the supply chain, as buyers operate at a zero profit margin. Most models in the literature assume that factor markets work well and that contracts are enforced, and ignore vertical coordination and specific contracting costs. These factors are important in reality. While factor markets work imperfectly in less developed regions, vertical coordination occurs in modern food chains all over the world, and search costs can be significant everywhere as well. These conditions not only have implications for the distribution of rents in food supply chains, but they also have an impact on price transmission.

The specific architecture of modern supply chains, which often involves vertical ties and requires crucial investments by buyers in contract-specific costs, implies that price transmission is discontinuous and depends on the nature and the amount of contracting costs. This obviously has important implications for empirical research in this area.

It also implies that the traditional logic, which assumes that weaker price transmission is associated with lower supplier welfare as powerful intermediaries in the supply chain are capturing all rents, is not universally applicable. In particular, we have shown that in vertically coordinated high-quality supply chains, conditions may arise under which farmers are better off in a context where price transmission is weaker, that is, the same factors that reduce price transmission cause a transfer of rents to farmers.

Acknowledgement

This research was funded by the European Commission (FP7 TRANSFOP project—<http://www.transfop.eu>), the Research Foundation Flanders (FWO) and KU Leuven (Methusalem Funding) and carried out before Anneleen Vandeplas joined the European Commission. Opinions expressed in this chapter are those of the authors and do not necessarily reflect the view of their institutions. The authors thank Steve McCorriston for guidance on the project; Rich Sexton for inspiration; Stefan von Cramon-Taubadel, Paolo Sckokai, Chema Gil, Ian Sheldon, Jim Vercammen, and seminar and conference participants in Leuven (LICOS), Washington (AAEA), Brussels (CEPS and TRANSFOP), and Toulouse (TRANSFOP) for useful comments; and Elena Briones Alonso and Elfriede Lecossois for help with graphs and editing.

References

Becker, G. S. (1962) Investment in human capital: a theoretical analysis. *Journal of Political Economy* 70(5): 9–49.

Bellemare, M. F. (2012) As you sow, so shall you reap: the welfare impacts of contract farming. *World Development* 40(7): 1418–34.

Birthal, P. S., P. K. Joshi, and A. Gulati (2005) *Vertical Coordination in High-Value Food Commodities: Implications for Smallholders* (MTID Discussion Paper No. 85). Washington DC: Markets, Trade and Institutions Division, International Food Policy Research Institute.

Bonnet, C. and V. Réquillart (2012) *Sugar policy reform, tax policy and price transmission in the soft drink industry*. TRANSFOP Working Paper No. 4, from the Transparency of Food Pricing Project under the EC Seventh Framework Programme (Grant Agreement No. KBBE-265601-4-TRANSFOP). From the Transparency of Food Pricing Project under the EC Seventh Framework Programme (Grant Agreement No. KBBE-265601-4-TRANSFOP). Brussels.

Bonnet, C. and S. B. Villas Boas (2013) *An analysis of asymmetric consumer price responses and asymmetric cost pass-through in the French coffee market*. TRANSFOP Working Paper No. 10, from the Transparency of Food Pricing Project under the EC Seventh Framework Programme (Grant Agreement No. KBBE-265601-4-TRANSFOP). Brussels.

Chang, H. and G. Griffith (1998) Examining long-run relationships between Australian beef prices. *Australian Journal of Agricultural Economics* 42: 369–87.

Crespi, J. M., T. L. Saitone, and R. J. Sexton (2012) Competition in U.S. farm product markets: do long-run incentives trump short-run market power? *Applied Economic Perspectives and Policy* 34(4): 669–95.

Davidson, J., A. Halunga, T. A. Lloyd, S. McCorriston, and C. W. Morgan (2012) *Explaining UK food price inflation* (TRANSFOP Working Paper 1). From the Transparency of Food Pricing Project under the EC Seventh Framework Programme (Grant Agreement No. KBBE-265601-4-TRANSFOP). Brussels.

Doyle, C. and R. Inderst (2007) Some economics on the treatment of buyer power in anti-trust. *European Competition Law Review* 28: 210–19.

Dries, L., E. Germenji, N. Noev, and J. Swinnen (2009) Farmers, vertical coordination, and the restructuring of dairy supply chains in Central and Eastern Europe. *World Development* 37(11): 1742–58.

European Commission (2009) A *Better Functioning Food Supply Chain in Europe*. Communication from the Commission to the European Parliament, the Council, the European Economic and Social Committee and the Committee of the Regions. Brussels: European Commission.

FAO (2009) *The State of Food Insecurity in the World 2009: Economic Crises—Impacts and Lessons Learned*. Rome: FAO.

Goodwin, B. and M. Holt (1999) Price transmission and asymmetric adjustment in the US beef sector. *American Journal of Agricultural Economics* 81: 630–7.

Gow, H. R., D. H. Streeter, and J. F. M. Swinnen (2000) How private contract enforcement mechanisms can succeed where public institutions fail: the case of Juhocukor a. s. *Agricultural Economics* 23(3): 253–65.

Headey, D. (2011) *Was the Global Food Crisis really a Crisis? Simulations versus Self-Reporting* (No. IFPRI Discussion Paper 01087). Washington DC: Development Strategy and Governance Division, International Food Policy Research Institute.

Holm, T., J.-P. Loy, and C. Steinhagen (2012) *Cost pass-through in differentiated product markets: a disaggregated study for milk and butter* (TRANSFOP Working Paper 6). from the Transparency of Food Pricing Project under the EC Seventh Framework Programme (Grant Agreement No. KBBE-265601-4-TRANSFOP). Brussels.

Inderst, R. and N. Mazzarotto (2008) Buyer power in distribution. In W. D. Collins (ed.), *Issues in Competition Law and Policy* (Vols. 1–3, Vol. 3). Chicago: American Bar Association Section of Antitrust Law.

Jacoby, H. G. (2013) *Food prices, wages, and welfare in rural India*. Policy Research Working Paper 6412. Washington DC:, World Bank.

Keefer, P. and S. Knack (2005) Social capital, social norms and the new institutional economics. In *Handbook of New Institutional Economics*. New York: Springer.

Klein, B. and K. B. Leffler (1981) The role of market forces in assuring contractual performance. *Journal of Political Economy* 89(4): 615–41.

Maertens, M. and J. Swinnen (2008) Standards as barriers and catalysts for trade, growth and poverty reduction. *Journal of International Agricultural Trade and Development* 4(1): 47–61.

McCorriston, S., C. W. Morgan, and A. J. Rayner (1998) Processing technology, market power and price transmission. *Journal of Agricultural Economics* 49(2): 185–201.

McCorriston, S., C. W. Morgan, and A. J. Rayner (2001) Price transmission: the interaction between market power and returns to scale. *European Review of Agricultural Economics* 28(2): 143–59.

McCorriston, S. and I. M. Sheldon (1996) The effects of vertical markets on trade policy reform. *Oxford Economic Papers* 48: 664–72.

Meyer, J. and S. von Cramon-Taubadel (2004) Asymmetric price transmission: a survey. *Journal of Agricultural Economics* 55(3): 581–611.

Minot, N. (2012) *Food price volatility in Africa: has it really increased?* IFPRI Discussion Paper No. 01239. Washington DC: International Food Policy Research Institute.

Minten, B., L. Randrianarison, and J. Swinnen (2009) Global retail chains and poor farmers: evidence from Madagascar. *World Development* 37(11): 1728–41.

Nash, J. (1953) Two-person cooperative games. *Econometrica: Journal of the Econometric Society* 21(1): 128–40.

Rapsomanikis, G. (2011) Price transmission and volatility spillovers in food markets. In A. Prakash (ed.), *Safeguarding Food Security in Volatile Global Markets*. Rome: Food and Agriculture Organization of the United Nations.

Sexton, R. J. (2012) Market power, misconceptions, and modern agricultural markets. *American Journal of Agricultural Economics* 95(2): 209–19.

Sharma, R. (2011) Review of changes in domestic cereal prices during the global price spikes. Rome: Food and Agriculture Organization of the United Nations.

Svejnar, J. (1986) Bargaining power, fear of disagreement, and wage settlements: theory and evidence from US industry. *Econometrica: Journal of the Econometric Society* 54 (5): 1055–78.

Swinnen, J. (2011) The right price of food. *Development Policy Review* 29(6): 667–88.

Swinnen, J., L. Knops, and K. Van Herck (2014) Food price volatility and EU policies. In P. Pinstrup-Andersen (ed.), *Food Price Policy in an Era of Market Instability*. Oxford: Oxford University Press.

Swinnen, J. and P. Squicciarini (2012) Mixed messages on prices and food security. *Science* 335(6067): 405–6.

Swinnen, J. and A. Vandeplas (2010) Market power and rents in global supply chains. *Agricultural Economics* 41: 109–20.

Swinnen, J. and A. Vandeplas (2011) Rich consumers and poor producers: quality and rent distribution in global value chains. *Journal of Globalization and Development* 2(2): 1–28.

Verpoorten, M., A. Arora, N. Stoop, and J. Swinnen (2013) Self-reported food insecurity in Africa during the food price crisis. *Food Policy* 39: 51–63.

von Cramon-Taubadel, S. (1998) Estimating asymmetric price transmission with the error-correcting representation. *European Review of Agricultural Economics* 25: 1–18.

Wang, X., H. Tadesse, and T. Rayner (2006) *Price transmission, market power and returns to scale: A note*. University of Nottingham Discussion Papers in Economics No. 06/07.

Weldegebriel, H. T. (2004) 'Imperfect price transmission: is market power really to blame?' *Journal of Agricultural Economics*, 55: 101–14.

Wohlgenant, M. K. (2001) Marketing margins: empirical analysis. In B. Gardner and G. Rausser (eds), *Handbook of Agricultural Economics* (Vol. 1). Amsterdam: Elsevier Science.

8

A Supply Chain Perspective on Price Formation in Agri-Food Chains

Gerhard Schiefer and Jivka Deiters

8.1 Introduction

Prices and price distributions along the food value chain are usually the result of the adoption of processes that build on a variety of indicators, including among others costs, market position, and the institutional and regulatory environment. In stable conditions, they can be considered as reflecting a market balance that is a 'fair' representation of actual conditions. In times of change this balance is distorted, raising the questions of how and how fast prices and price distributions along the food chain could adapt towards reaching a new balance that fits the emerging situation. In this chapter, we address the issue of price transparency in food chains from a supply chain perspective. The methodology applied is very different to the economists' approach to pricing issues in supply chains in that we highlight the main features of a supply chain perspective. The supply chain methodology draws upon case studies, creating scenarios, expert focus groups, and the role of networking and from this analysis, highlights priorities that participants in the food chain can adopt to respond to specific pressures facing agri-food chains. We draw on these features to illustrate the insights that can arise from a supply chain perspective and, in the course of the discussion, highlight these issues with reference to the sustainability issues that are receiving increasing attention from stakeholders and consumers. Since sustainability issues involve social, environmental, and economic dimensions, responding to the sustainability challenge may have a profound effect on the organization of food chains, the relationships between enterprises within and across different stages of food supply chains, and, related to this, the distribution of costs and returns (as reflected in price developments) throughout the supply chain.

In this chapter, we first provide an overview of the relevance of the supply chain approach in dealing with chain relationships, as well as its relevance for policy and stakeholders in enterprises and the institutional environment of enterprise activity. This discussion provides the basis for a more in-depth characterization of the present sustainability challenge and the limitations in food chain organizations to meeting these challenges, as well as to appropriately adjusting their relationships and determining price developments in a time frame limited by pressures from society and markets. Subsequent sections focus on developing food chain transparency; identifying a chain-encompassing view on so-called 'hot spots' for best adapting to emerging sustainability challenges; the potential for regaining investment costs through an increase in market prices; and engaging in new and emerging forms of horizontal and vertical networking for supporting adaption to market challenges, for realizing efficiency gains for covering investment costs, and for assuring an appropriate distribution of efficiency gains and market appreciation, for example, for realizing an appropriate price formation along the chain.

8.2 Supply Chain View and its Policy Relevance

The supply chain view aims to capture an accurate picture of the situation in individual supply chains. This differs from a statistical approach which (assuming the availability of an appropriate database), while it can provide a 'representative' picture, nevertheless levels out differences between individual cases; the risk is that the 'true' picture of the situation in individual cases gets lost. The supply chain view is most relevant if a sector and its supply chains are faced with challenges that are different from experiences in the past, reducing the value of available data for analysis. It allows the researcher to draw a picture of supply chain reality which helps explain specific conditions, and to reach conclusions on supply chains' expected or possible reactions upon meeting the challenges they face.

In this context, a supply chain view builds on the identification of possible future scenarios and focuses the analysis on supply chains that represent certain conditions but have a broader relevance beyond the individual cases. Depending on interest, the understanding of 'relevance' could refer to different types of representation. As examples, relevance could refer to capturing a broad range of existing supply chains or, alternatively, to capturing the situation in most innovative chains that are advanced in dealing with the challenges of the future. In this chapter, we focus on the second alternative, which provides support in understanding developments in the future.

A Supply Chain Perspective on Price Formation in Agri-Food Chains

The analysis of supply chains involves consideration of specific features including:

a) relationships between its members;
b) the reasoning of members, which determines their decision-making behaviour; and
c) the legal and institutional environment in which they operate.

The extent of transparency within a specific supply chain plays a crucial role in understanding supply chain relationships and behaviour. Concerted action towards meeting emerging challenges, as well as the understanding of resulting consequences in terms of investments, costs, prices, and price distribution, all depend on the level of transparency within the chain and towards consumers. This underpins the role of case studies in the supply chain methodology. As each case is in principle different from others, it is part of the research effort to judge which conclusions can be drawn that reach beyond the detailed case study to have wider relevance. The judgements are principally reached using an evidence-based reasoning approach. One needs to be aware that those judgements are 'soft' results, where assessors need to judge for themselves to what extent they are prepared to follow the arguments and accept the conclusions.

Results from a supply chain view may support enterprises in supply chain development involving optimization of chain relationships and better adaptation to meeting the emerging challenges. They may also contribute to supporting policy in sector activities. Policy intervention may focus on supporting enterprises and chains in meeting the challenges ahead or provide guidance towards a scenario preferred by society. In this context, policy has the option of influencing stakeholders' reasoning or the legal and institutional environment in which reasoning and enterprise activities take place.

8.3 Realizing a Supply Chain View towards Price Formation in Sustainability Scenarios

8.3.1 *The Role of Scenarios*

With the many developments and pressures linked with assuring the sustainability of the food sector in its ability to provide food that is safe, readily available, affordable, and of the quality and diversity consumers expect (CIAA, 2007, 2014), any discussions on initiatives requires a delineation of focused scenarios. The scenarios describe artificial alternative futures which focus on disruptive alternatives.

The key phrase is 'alternative futures'. This distinguishes scenarios from forecasts which relate to a most probable future. However, they are still based on an analysis of different developments and are, as such, more definite than pure speculation. There have been a substantial number of efforts to formulate scenarios (Rabbinge and Linnemann, 2009). As an example, one might focus on the regularly updated initial scenario study by the Standing Committee on Agricultural Research (SCAR, 2009, 2011). This builds on a 'baseline scenario' which roughly describes a continuation of observed trends and developments. The development of this approach involves alternative scenarios compared to the baseline, each one characterized by a specified and distinct disruption of the baseline such as 'climate shock', 'energy crises', 'food crises', and 'cooperation with nature'.

The delineation of different scenarios is based on the consideration of a number of drivers such as societal and demographic changes, the macro-economy and trade, climate change and global warming, environment, energy, science and technology, and health. The assessment of different drivers is approached through the control of input or output characteristics that have a critical impact on them. Characteristics that have a positive impact on the drivers are those that may add value to products, which in turn might be appreciated by markets. As an example, carbon emissions of food production have an impact on the driver 'climate change', which in turn is linked to the possible scenario 'climate shock'. To summarize, we are dealing with a breakdown of scenarios to, first, drivers and, second, impact characteristics.

In the context of responding to the challenge of sustainability in food supply chains, we can identify four major drivers. These are: (a) 'reduction of waste' which may contribute to an increase in food availability, which in turn may be linked to the driver 'societal and demographic changes';[1] (b) 'reduction in GHG emissions' including carbon footprint may be linked to the driver 'climate change and global warming';[2] (c) the characteristic 'reduction in water use' represented by water footprint may be linked to the driver 'environment';[3]

---

[1] Waste may be identified as food which is aimed at but unfit for human consumption. According to FAO (2011), about one-third of the edible parts of all food produced for human consumption becomes loss or waste. Food losses occur at all stages of the chain (Parfitt et al., 2010). In Europe and North America, losses mainly occur at the end of the supply chain at retail and consumer households, while in developing societies, the production of waste is more closely linked to early stages of the chain.

[2] For emissions, the Intergovernmental Panel on Climate Change (IPCC) provides a clear specification. It concludes that anthropogenic greenhouse gas (GHG) emissions (carbon dioxide, methane, nitrous oxide, and halocarbons) have been responsible for most of the observed temperature increase since the middle of the twentieth century (IPCC, 2007). A 2006 study for the European Commission found that food accounts for 31% of the EU-25's total GHG impacts (Eder and Delgado, 2006). GHG's contribution to global warming is expressed in CO_2 equivalents (carbon footprint). CO_2 is rated with a factor of 1, nitrous oxide with a factor of 298, and methane with a factor of 25 (Brander, 2012).

[3] About 70% of freshwater consumption is due to agricultural production (FAO, 2009). As a consequence, any water shortages endanger food production. The problem is further aggravated as future climate changes may lead to a 20% increase in water shortage (UN, 2006: 46). Water

A Supply Chain Perspective on Price Formation in Agri-Food Chains

and (d) the characteristic 'reduction in energy use' referring to non-renewable energy sources is directly linked to the respective driver.[4] All of these issues affect all stages of the food supply chain. Additional characteristics with limited chain relevance are not part of the applied case study. They include, among others, animal welfare (which are concentrated on the farm level only), fair trade, and GMO-free product lines (the challenge in GMOs relating to the organization of parallel logistics chains, which might reduce some of the economies of scale but not change the relationships between stages of the food chain).

An earlier study (Amani and Schiefer, 2012) provides a literature overview of our present knowledge (mostly based on case study analysis) on the impact of various stages of food chains in different product lines on value adding characteristics, especially on GHG emissions and energy. The impact of value adding characteristics along the food chain may differ substantially between products. As an example, one might look at the differences in the use of energy in the production of fresh meat or hamburger. In fresh meat, almost all energy use is concentrated in agriculture, whereas in hamburger production, the use is more evenly distributed among agriculture, bakery, and packaging (Amani and Schiefer, 2012).

The key in the analysis of adding value characteristics and their linkage with investments, costs, and pricing is the identification of a baseline from which to start the analysis. A suitable approach could build on the common characteristics of commodities or, in non-commodity products, on what is commonly considered to represent the basic expectations or the present 'average'. While the identification of a basic expectation level might look complex from a theoretical point of view, it is a feasible working base building on very 'obvious' expectations such as food safety guarantees, the exclusion of child labour, the exclusion of obvious cruelty to animals, the avoidance of obvious damage to the living environment, and so on.

8.3.2 Identifying Relationships between Enterprises along the Chain

Relationships between enterprises along the chain can be analysed in various dimensions. For understanding supply chain problems in the food sector, the following dimensions of analysis provide specific insights. They involve drivers for food chain development that are decisive for adapting to emerging challenges and for supporting appropriate considerations of investments,

---

consumption in the production of goods is sometimes expressed in terms of 'water footprints' (<http://www.waterfootprint.org>).

[4] Energy is an input factor all along the food chain. It is used as fuel or electricity in the utilization of machinery, in heating, cooling, lightning, but also in the production of fertilizer or pesticides.

171

costs, prices, and price distributions. They involve (a) the resource-based view and (b) the organizational view, both of which we outline below.

8.3.2.1 THE RESOURCE-BASED VIEW

The relationships in chains are to some extent dependent on the relevance of their contribution to the production, distribution, and sales of products. In times of overproduction of agricultural products, the standing of farms has been quite weak. However, times change; resources provided by farms are increasingly scarce and will subsequently strengthen their role in chain relationships. This refers to classical production resources such as land, but also to quality and sustainability characteristics of agricultural production. There are examples which demonstrate this change. One of them, discussed in more detail below, involves a large retail chain that was required to provide substantial incentives to muster enough farms that could support its quality and sustainability sales programme.

Increasing dependency in assuring quality and sustainability characteristics along the whole chain supports a balanced status of stakeholders in the chain and favours, in turn, the development of vertical networking activities. *Vertical networking* with appropriate balancing of interests is an approach that supports the development of stable relationships in times of increasing dependency between production, distribution, and sales. Farms can more easily adapt to the changing expectations of markets, while retail can secure its supply for meeting market needs.

8.3.2.2 THE ORGANIZATIONAL VIEW

For realizing the appropriate balancing of interests in supply chains, relations need to be based on an appropriate organizational structure. The structure of food chains sets them apart from chains in other production sectors. In general, both ends of the chain, from farm supply to retail, build on large and even internationally active enterprises, while those enterprises in between, such as agriculture and to some extent processing companies, are predominantly small and medium-sized enterprises (SMEs) (CIAA, 2007). This causes problems for both the large companies and the SMEs. Large companies have problems in sourcing and assuring conformity in quality and product characteristics, while small companies are limited in their price negotiation ability.

This is the basis for horizontal networking initiatives among SMEs, which support efficiency towards their suppliers and customers but, in addition, support SMEs in negotiating their appropriate share in the balancing of costs and returns in chains. *Horizontal networking* allows SMEs to realize their potential strength in chain relationships. As an alternative to networking, SME companies might merge and grow. In a case study chain example reported

A Supply Chain Perspective on Price Formation in Agri-Food Chains

in Fritzen and Schiefer (2013), farms improved their standing through horizontal networking while the receiving trade company chose to merge with others and to grow in size, the options pursued by the different participants depending on market opportunities and management interests.

In summary, integrating the tendencies evolving from the resource-based view and the organizational view, food chain relationships that support developments towards sustainability will be characterized by an intensification of horizontal and vertical networking relationships. Both developments will have a profound effect not just on the negotiation situation regarding production characteristics, but also on price formation and price distribution along the chain.

8.3.3 Efficiency Issues

Improvements in sustainability of enterprises and food chains build on improvements in individual sustainability characteristics of operations and products, which constitute the *added value* of products. Such improvements may require investment and operational changes in enterprises along the food supply chain. They might be feasible to a certain extent without costs or might even involve cost reductions. However, beyond a certain point, improvements in sustainability characteristics in the production and distribution of food lead to an increase in costs.

For enterprises and food chains as a whole, this poses a challenge which differs from challenges of the past, where the focus was primarily on production and on improvements in efficiency. In principle, improvements in efficiency have a positive effect on costs, which drives developments without outside interference. Efforts towards improvements in environmental and social sustainability cannot build, in general, on such clear relationships with costs, which constitutes a conflict in reaching sustainability along the chain. This raises the question of how to balance costs linked with improvements in environmental and social sustainability.

One could distinguish between the following two scenarios: first, if improvements in sustainability characteristics (value characteristics) allow the realization of price premiums with consumers, the challenge is the *distribution of market gains* along the chain in line with the increase in costs; second, if improvements in sustainability characteristics (value characteristics) do not or only partly allow the realization of price premiums with consumers, enterprises are challenged to improve in efficiency, organization, and management in order to offset some of the additional costs and to *distribute efficiency gains* according to engagements in sustainability investments.

In dealing with the different issues, one needs to consider prices and costs simultaneously. This is captured in the profit margin. While the identification of sales prices along the chain and towards consumers is a straightforward task, the identification of costs is difficult to reach. There is a lack of appropriate statistical data. Furthermore, in multi-product enterprises such as retail, the calculation of costs for individual products does, apart from the purchase price, very much build on the allocation of indirect and overhead costs. Information about costs is also a sensitive issue with enterprises, which are reluctant to provide appropriate information for suppliers and customers in the chain and, in consequence, in case study analysis.

8.3.4 The Role of Transparency

Discussions on changes in value adding characteristics of products and processes and the need for balancing costs and returns in case of an increase in costs depend on *transparency* in (a) prices, and especially price premiums at the consumer end of the supply chain, (b) enterprises' cost engagement towards improvements in sustainability characteristics, and (c) potential gains in efficiency improvements along the chain.

In this context, transparency is of relevance both for enterprises along the chain and for consumers. Transparency with regard to enterprises supports them in improving on critical food characteristics that are of major relevance to consumers and society (Schiefer and Deiters, 2013b). Transparency is the basis for a readjustment in the distribution of profit margins and is the precondition for individual enterprises to invest, if not forced to through regulations or market pressures. The mutual information exchange on costs, returns, and the willingness to balance the distribution of costs and returns is a precondition for swiftly moving towards increased sustainability.

Transparency towards consumers may allow them to better identify the value adding characteristics and make informed decisions that fit their needs. Transparency builds on appropriate signals which integrate available information and communicate a certain message to recipients (e.g., 'food is safe'). One of the basic signals is price. However, their message is blurred. Higher prices communicate a message of being of value to customers or of being more valuable than competing products. The value is defined by the expectations which customers link with the product. Pricing along the chain could reflect the distribution of production and distribution costs along the chain or include, additionally, the contribution to realizing and communicating the value adding characteristics to consumers. While the first part could be clearly quantified, the second part depends on the evaluation of customers, which might differ between customers from different backgrounds but also change over time.

Realizing transparency is one of the most complex and fuzzy issues the food supply chain faces. The challenges relate to complexities in food products and processes but are also due to the dynamically changing open network organization of the food sector with its multitude of SMEs, its cultural diversity, differences in expectations and the ability to serve transparency needs, and its lack of a consistent appropriate institutional infrastructure that could support coordinated initiatives towards higher levels of transparency throughout the food value chain (Schiefer and Deiters, 2013b).

There are a substantial number of initiatives working to improve transparency. These include initiatives that (a) deal with the establishment of an appropriate communication *infrastructure*, (b) focus on the identification and communication of appropriate *indicators*, (c) deal with the identification and realization of system *functionalities* that serve information needs in specific scenarios, and (d) transform information into appropriate *messages* at the consumer end. These initiatives demonstrate the broad range of issues that need to be dealt with but also the many deficiencies in the realization of transparency on value added characteristics. An overview on deficiencies is provided in the Strategic Research Agenda of the European project *Transparent_Food* (Schiefer and Deiters, 2013b).

8.4 Improvement Potentials ('Hot Spots')

There is a broad literature on the relevance of sustainability characteristics in different stages of chains related to different production lines (Amani and Schiefer, 2012). However, knowledge of impact is different from knowledge about potential opportunities for changes in impacts and the related consequences for enterprises and the chain in terms of costs and returns. The scope for addressing these opportunities, so-called 'hot spots', includes drawing on expert focus groups.

The intensity of negative impacts of production and distribution activities in enterprises along the chain that arise from responding to sustainability issues may not match the improvement potential within a given production and distribution structure. As an example, the negative impact of milk production on GHG emissions of farms is primarily due to the emissions of the cow population. If an end of milk production is not an option, the potential for a reduction of GHG emissions is very limited. However, sustainability can be addressed throughout the supply chain as a whole, covering GHG emissions, water, energy, and waste (see Fritzen et al., 2012). Taking as an example the savings potential in the grain supply chain, savings could relate to energy savings (primarily in industry, bakeries, and to some extent in farms' drying processes); waste reduction (primarily at retail and in food manufacturing, e.g., through changes in packaging); and water savings, primarily in industry (e.g., cleaning processes), with limited potential for GHG emission reduction (to some extent at farms through fuel reduction).

These developments would affect the distribution of profit margins along the chain, with the processing stage at a disadvantage. Within the chain, the greatest potential for savings is in industry, despite the fact that agriculture has a major impact on some of the characteristics. The picture in other product lines is quite similar. This places special responsibility on industry and retail regarding investments towards improvements in sustainability, while all stages—including agriculture—would have to cooperate in regaining an appropriate balance in the distribution of costs and returns.

8.5 The Role of Networking for Price Formation in Times of Change

8.5.1 *The Network Focus*

Networking as a means for meeting emerging challenges for food chains and for improving the balancing of individual returns in line with the engagement in terms of investments and costs for individual members of the food chain has been referred to in Section 8.5.1. The classical network arrangements focus on either horizontal cooperation or vertical cooperation. Typical examples of horizontal cooperation are traditional cooperatives. Horizontal networking allows improvements in SMEs' standing within a chain and support investments in coping with external pressures through efficiency gains in production and marketing. They bring individual enterprises together for joint marketing activities, building on some common agreements in production and process organization. Horizontal networking, especially in agriculture, has been an established approach to dealing with emerging challenges with which individual farms could not cope (Schiefer and Deiters, 2013a).

Vertical cooperation relates specifically to the linkages between stages in the food chain. Vertical networking allows chains to reach a production and distribution level that better matches the emerging needs of consumers and society, especially regarding issues of sustainability, while at the same time assuring a new balance in the distribution of costs and returns. Food chain cooperation may be linked to improvements in marketing that build on commonalities along the whole chain of enterprises involving the branding of products. Food chain cooperation might also be linked to improvements in risk management, assuring appropriate controls along the chain, specifying agreements on production quantity (e.g., through contracts), and assuring a predetermined quality at the various stages of the chain. Examples are the classical organic production chains or chains that build on contract farming.

A Supply Chain Perspective on Price Formation in Agri-Food Chains

As the vast majority of enterprises in the food sector (CIAA, 2007, 2014) as well as in farming are SMEs, the question of the value of networking in moving towards increased sustainability is of major relevance for the sector's development in that direction. Networking may provide the necessary motivation and the necessary market relevance to induce enterprises to make investments towards increased sustainability.

Networking is especially relevant for SMEs, because the individual companies usually do not have the internal and external strength to move ahead and gain market recognition. In addition, networking could support the interaction and knowledge exchange deemed necessary for a successful move into new and emerging trends and requirements in food supply (Schiefer and Deiters, 2013a). With increasing pressure towards improvements in sustainability, the classical network arrangements which constitute baseline cooperation options are being further developed in order to better meet the challenges towards sustainability and the balancing of interests (including costs and prices) in times of change.

8.5.2 Emerging Network Alternatives

In discussions with experts and networks engaged in sustainability developments, a number of emerging network alternatives were identified that could be structured as follows: (a) production driven networks covering open horizontal/vertical networks and dedicated horizontal/vertical networks; (b) customer driven networks covering open horizontal networks and dedicated horizontal integration; and (c) complete horizontal integration. Some of these networks evolved from networks already in place, while others were specifically initiated for dealing with sustainability issues. For example, open networks represent networks that are not linked to any specific chain of supplier/customer relationships (chain). They were usually initiated with food safety and food quality in mind, and later broadened their scope towards environmental and social issues.

The analysis in this chapter builds on case studies of emerging networks, each case study representing a typical network design aimed at better meeting emerging challenges and supported by feedback from an expert group of representatives from retail and industry linked to various product lines and a consumer focus group. The focus of the case study analysis is on identifying the possibility of reaching price premiums for sustainability products, the effect of consumer communication on their appreciation of sustainability products, the supply chain's ability and willingness to assure balancing of costs and returns along the chain, and opportunities for supporting developments towards sustainability through networking. The network case studies are of specific relevance as they allow discussion of opportunities for coordinating improvements towards sustainability among many SMEs involved in a chain, gaining in efficiency that could offset sustainability improvement costs, and supporting the balancing of costs and returns for SMEs in a chain.

For evaluating the effect of engagements in value adding characteristics and of networking on the distribution of prices, costs, and eventually profit margins along the chain, the case study approach takes the present distribution as a base and focuses on changes that may result from engagements in value adding characteristics and networking. This is based on the argument that the present distribution of profit margins along the chain is the result of an ongoing continuous adjustment process in the market. It is interesting to note that various case studies discussed in Fritzen et al. (2012) suggest that the profit margin per product unit is about the same in enterprises along the chain, with a somewhat higher margin for agriculture which, however, may vary considerably between years depending on harvests and demands.

The selected examples of alternative network arrangements given below, drawing on Fritzen and Schiefer (2013), are all active in promoting sustainability developments covering a broad range of issues, including environmental, social, and ethical issues. As such they are forerunners of developments that could be envisaged for the future. The *GlobalG.A.P* network (<http://www.globalgap.org>) is a global *network of farm* enterprises that follow similar rules in the organization and control of their production processes. They are viewed as a network of their own; for example, some retailers only accept products from the GlobalG.A.P network. According to our classification, the network is a *customer driven network* as it was initiated by retail groups. It is an *open horizontal network* that farms everywhere can join. The *Q&S (Quality and Safety)* network (<http://www.q-s.de/home_gb.html>) represents a *vertical network of companies* (or horizontal networks of companies) *at all stages of the food chain* that at each stage and within a certain food sector follow similar rules in the organization and control of their business processes. It evolved out of the farming community and was extended step by step to later stages of the food chain. Enterprises at all stages are considered as being members of a network, as some customers and eventually some retail groups accept products from the Q&S network. According to our classification, the network is an *open horizontal/vertical and production driven network* that could be joined by any company anywhere that follows its rules. The *Naturland* network (<http://www.naturland.de/welcome.html>) represents a *vertical network of companies* (or horizontal networks of companies) *at all stages of the food chain* including farms, processing industry, and marketing groups that cooperate in providing the 'Naturland Brand', a major brand in the organic sector. It evolved out of the production sector, representing a *production driven network*. The members follow dedicated rules that guarantee the brand image. Products are sold in

A Supply Chain Perspective on Price Formation in Agri-Food Chains

dedicated retail outlets that are also members of the network but also in the general market. According to our classification, the network is a *dedicated network* with *horizontal and vertical cooperation*. An example of a retail network which is currently being developed is a *customer driven network* involving a major international retail group with headquarters in Germany that is launching a sustainable product line (Fritzen and Schiefer, 2013). For serving this product line, the group initiates dedicated vertical network building, especially on selected farms (horizontal network) but also involving other enterprises along the chain that may be part of horizontal networks as well (*horizontal/vertical network*) and promotes production of products fitting their requirements throughout Europe. At the other end of the supply chain, an example of a producer network (currently under development) involves a major cooperative in France that builds on a farming community but also includes a processing industry that is the market leader in some segments (Fritzen and Schiefer, 2013). It is working to launch a sustainable product line with a dedicated network of farmers and processing companies (*horizontal/vertical network*). Retail is not part of the network. According to our classification, the network is a *production driven network*.

8.6 Networking Case Experiences

All networks claim efficiency gains. *Naturland* (organic production) and the retail network claim the realization of possible price premiums. Both claim that price premiums are being channelled back to stages where improvements in sustainability result in additional costs. Apart from the production network, all networks involve the retail stage, which they claim provides them with a competitive advantage at the retail stage.

The principal organization of all networks is quite similar. Farms, and to some extent other members of the chain, follow certain production and distribution rules that are formulated and controlled by an institutional unit. Major differences concern the consideration of sustainability characteristics in the rules and governance of the institutional unit, which may be linked to the production stage, the retail stage, or served by all members of the chain.

With a broad base in the farming community, the *GlobalG.A.P.* and *Q&S* networks are mostly directed towards business customers and have limited consumer communication. They provide an advanced baseline level in food safety and food quality as well as in environmental and social concerns. This is underlined by the fact that the products do not reap higher sales prices but are more or less requirements for market access. Market access and improvements in organizational and managerial efficiency towards savings in costs are the major benefits associated with the networks. *Q&S* may gain a competitive (price) advantage in sectors involving processing, as it can provide guarantees for all claims regarding quality or sustainability throughout the chain. Combining *GlobalG.A.P.*'s global sourcing base with *Q&S*'s value chain controls would match the advantages of both networks.

Products from *Naturland* and the retail network are both visibly present to consumers in retail outlets. Supporting visibility is part of both networks' marketing policy. They both cover all stages of the chain and can provide guarantees for claims associated with the production and distribution of products. As a result, their products are clearly differentiated from others, allowing them to potentially reap a price premium. The added value is mostly channelled back to where additional costs occur but may not cover them completely. The linkage of sustainability with additional drivers for sale such as 'regional' may close the gap, especially if market stability and benefits from organizational efficiency as well as from inherent advisory services are included in the calculation. The retail network has a competitive advantage in reaching broad market acceptance as it is part of a group's strategic development programme. This may be offset by *Naturland*'s focus on a dedicated target group.

Matching the production network against the model represented by the retail network, it is doubtful that the production network will receive broad market acceptance if not dedicatedly supported by a retail group. If it is not, it may create and maintain a niche brand with a limited sourcing base. In this scenario, it is difficult to see room for price benefits at the retail stage.

8.6.1 *Advanced Networking Organization*

The retail network represents the most advanced networking approach as it supports horizontal networking at all stages of the chain while integrating all stages into a vertical network design. Its focus is on improving efficiency as well as on sustainability characteristics aimed at creating a 'sustainability brand' for the mainstream market, with the ambition of reaching a price premium even in a competitive situation. The approach was developed out of the *Global G.A.P.* model, as it is governed by retail but does not limit itself to farms, but rather integrates all stages of the chain.

In this network, all stages involved in the production and distribution of a certain product are evaluated according to their performance vis-à-vis a number of sustainability characteristics, such as energy, GHG emissions, water use, land use, waste emissions to water, and so on. The individual evaluations regarding the sustainability characteristics are aggregated in an 'ecological' index for the product for each stage. The aggregation across all stages provides an ecological index for the final product. Products which reach a certain threshold are integrated into the sustainability brand. Price premiums for

A Supply Chain Perspective on Price Formation in Agri-Food Chains

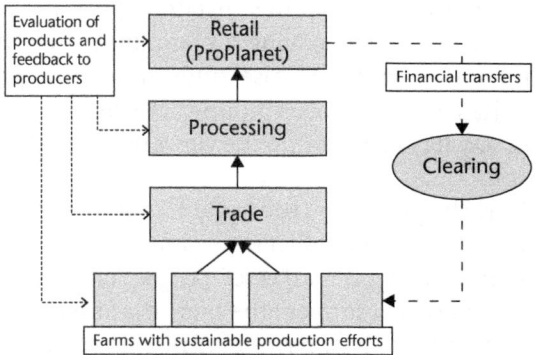

Figure 8.1. The retail network approach

the brand are at least partly channelled back to farms, which have to deal with an increase in costs through a clearing unit. The organizational set-up is outlined in Figure 8.1.

This network is characterized by a number of advanced features. Specifically, the network utilizes a broad range of sustainability criteria integrated into the sustainability index. The use of an index allows farms and other enterprises flexibility in developing towards sustainability goals and caters for different backgrounds and scenarios. The evaluation of products covers all stages of the chain so that a 'true' chain-encompassing sustainability index can be identified. Participants can reap financial benefits through a financial clearing system that transfers a part of sales premiums back to participants according to their level of engagement. Finally, evaluations are carried out under the supervision of an external expert board independent of retail and participants.

An interesting example of how profit margins may be distributed across the chain is represented in a chain consisting of farms, a trader, and retail (Fritzen and Schiefer, 2013). Requirements from retail participants regarding improvements in sustainability characteristics resulted in additional costs to farms that could not be covered by appropriate premiums in consumer prices. This led to two developments towards improvements in efficiency in farms and the trading stage that could cover additional costs. First, farmers entered into a network cooperation that covered part of the additional costs. Second, the trader paid a premium to farmers that covered the remaining cost deficiency while increasing its own efficiency to cover these increased premiums.

8.6.2 Regaining Costs through Consumer Price Increases

As it is assumed that an improvement in value adding characteristics is in the general interest of society but specifically of consumers, one might ask if the additional costs of investments towards sustainability could, at least to some extent, be carried by consumers paying a premium price. This question has been raised in interviews with the case study industry representatives and discussed within the consumer focus group that was comprised of people of different ages and sex. It is striking that the views between the different groups matched quite well.

Examples such as the one from the organic network (*Naturland*) demonstrate that investment costs can in fact be regained if the supply of products is limited and if there is a niche market of people able and willing to pay a premium. The examples also support the view that investment initiatives at individual stages and especially at farms may not lead to a major distortion in the distribution of profit margins along the chain, but that additional returns from consumer sales are to a major extent channelled back to farms as the stage where the major additional costs occur for delivery of the added value characteristics of organic chains.

The question arises as to what extent the situation in the market for organic products might match the situation in markets for products based on production and distribution processes with improvements in sustainability. It is evident from expert analysis that in the competitive environment of mainstream markets, additional costs cannot be regained through retail price increases as long as competitive products with lower production costs are on the market. Consequently, an investment in added value characteristics would need to be covered as much as possible by savings through improvements in process efficiency and management.

It was argued that consumers have little knowledge of the content of the term 'sustainability'. This is supported by a recent business study focused on Germany (Nestlé, 2011) that found that even if the term 'sustainability' is known by a majority of consumers, only about half of them can make sense of it. Sustainability is an argument which is not currently of market relevance and which may not become relevant soon. As a result, it is individual sustainability domains such as animal welfare, non-GMO, contamination, or, increasingly, social issues that determine the discussion at the retail end. This is in line with the results of the Nestlé (2011) study where consumers could phrase their importance rankings and willingness to pay for individual dimensions of sustainability. While importance rankings ranged from about 30 to almost 70 per cent, there was even a 'willingness to pay' of between 10 and more than 30 per cent, signalling an understanding of issues.

However, it was made clear in discussions that in the long run, limiting the focus to individual sustainability characteristics is not the way to go. The relevance of characteristics may change and more individual characteristics are becoming of interest. Furthermore, the extent to which individual sustainability characteristics are dealt with may differ between stages of the chain, making it almost impossible to communicate with consumers. This has been the argument behind the classical certification schemes as well. As a consequence, enterprises are looking for ways to appropriately communicate 'sustainability' to consumers. It is against this background that enterprises look for opportunities to link 'sustainability' with other indicators. These include linking 'sustainability' with 'organic', as consumers who are aware of sustainability characteristics tend to purchase organic food, considering it sufficient for environmental and other concerns. Another example is linking 'regional' with 'organic'. According to Nestlé (2011), about 37 per cent of all consumers buy local products on a regular base, while the number for organic products is only 13 per cent. Consumers associate local products with freshness, support of the local economy, short transportation distances, and knowledge about the origin. In this context, experts emphasized that regional production, in the view of consumers, involves a multitude of additional 'goodies' such as trustworthy claims, acceptable animal welfare considerations by responsible people, non-GMO, and so on. According to DBV (2011), organic food is mainly bought on the basis of a perceived higher health benefit, to some extent combined with understanding of animal welfare, environmental protection, and better taste.

Summing up the arguments, it can be stated that the following preference scale for the general public prioritizes regional over organic and organic over sustainable. This makes the combination of 'sustainability' and 'regional' the most attractive option for offering sustainably produced products to consumers. A similar opportunity to achieve acceptance of a premium for adding value characteristics may develop through a combination of sustainability characteristics with other attributes that promise an individual added value, such as health support as linked with organic products or the identification of certain production sites as an indicator for food safety and quality.

A case in point is the example of the sustainability initiative of the company Barilla. The company developed a double pyramid model that links healthy eating with sustainable production (Barilla, 2014). The well-known pyramid of eating recommendations with vegetables at the bottom (broad base) and meat at the top (limited consumption) is set in parallel with a reverse pyramid on negative environmental impacts, with plant production at the bottom (little negative impact) and animal production at the top (large impact). This double pyramid communicates the message that what is good for one's health is also good for the environment.

The line of discussion from expert analysis can be summarized as follows. First, if consumers have an alternative, the majority will not be prepared to pay an extra premium. They might prefer such products against 'traditional' products as long as prices are comparable, providing 'sustainable' products with a competitive advantage. Second, a small segment of people would be prepared to pay an extra premium, as is the case for organic products. It is assumed that the price premium would be somewhere between the price for traditional products and the 'reference price' for organic products. Prices for organic products would constitute the borderline, as 'organic' in the view of consumers involves a health premium not included in the 'sustainability' characteristics.

It is further argued that even for the segment of the population who would be prepared to pay an extra premium, those people need to be guided to the extra premium and get used to appropriate appreciation of the product. A case in point is an experiment carried out by a major meat processing company in Germany which tested consumer reactions on price premiums for poultry meat. Introducing poultry meat certified as 'Label Rouge (LR)', with its higher levels of sustainability standards in higher-level retail stores, was a failure with customers. With a switch from 'Label Rouge' to the label 'Nature and Respect' with a reduced price premium, the turnover increased. It became apparent that consumers who value certain sustainability aspects were not yet ready to pay a premium they considered too high. They would prefer to compromise and to focus on products with some improvements but with a lower 'price shock'. Moving towards increased sustainability may require a long process involving gradual improvements that consumers can digest.

8.7 Conclusion

To gain insights on price formation in food chains, this chapter has outlined the perspective from a supply chain approach, the key features of this approach being the development of scenarios, use of case studies, reference to focus groups, and expert opinion. We have highlighted this approach by addressing pricing issues in the supply chain and transparency in food supply chains more generally with reference to the issue of sustainability. However, while the emphasis has been on pricing issues, in addressing developments in food supply chain, prices are just one side of the coin. They have to be linked to costs and eventually the profit margin for products at the various stages of the food chain. In this analysis, we take the present distribution of the profit margins along the chain as a base from which to proceed. Our focus is on developments towards sustainability that would require investments and additional costs that are not evenly distributed along the chain. As the balance of costs and returns is the result of adjustment processes over time, transparency in costs, prices, and investment needs to support a rebalancing of costs and returns in line with developments towards improvements in sustainability.

The additional costs could be gained through a price premium with consumers and/or efficiency gains along the chain. However, price premiums are the exception and are found mainly in niche markets such as the market for organic products. In these cases, price premiums were in principle channelled back to the stages where the additional costs for reaching the higher level of sustainability occurred. In general, however, sustainability improvements do not pay off in terms of price premiums. This requires improvements in efficiency for reaching feasibility. Various case studies demonstrate that gains in efficiency are channelled to those stages in the chain where additional costs from sustainability initiatives occur.

For small and medium-sized enterprises, including farms and the majority of enterprises in the food industry, networking provides a major potential for realizing efficiency gains, not just in individual enterprises but also in the integration into the chain. There are different alternatives for how to organize networking. They evolve out of the need to meet the emerging challenges and to provide a platform on which an appropriate balance of costs and process can be reached. The most promising are the ones that build on a combination of horizontal and vertical networking, integrating all stages and reaching from farms to retail.

Based on these experiences, one could draw the conclusion that both the appropriate adaption of the sector to changing environments and the appropriate and short-term balancing of costs and returns along the chain in times of change is being supported by horizontal and vertical networking initiatives. As a consequence, policy initiatives in support of appropriate balancing of costs and prices in times of change might focus on facilitating networking initiatives and on removing obstacles for cooperation that may exist, for example, in opportunities for price policy arrangements within chains.

References

Amani, P. and G. Schiefer (2012). Future Scenarios of Regulatory Environments and Opportunities for Improving on Value Added Food Attributes in Food Chains. Report D5.1 of the EU project TRANSFOP to the European Commission, Brussels.

Barilla (2014). Available at <http://www.barillagroup.com/corporate/en/home/our-sustainable-model/sustainable-business-reports.html>.

Brander, M. (2012). Greenhouse Gases, CO2, CO2e, and Carbon: What do all these terms mean? Ecometrica. Accessed 15 November 2014, available at <http://ecometrica.com/white-papers/greenhouse-gases-co2-co2e-and-carbon-what-do-all-these-terms-mean/page/1/?filter&filter_category=&filter_date=&filter_topic=carbon>.

CIAA (2007). Strategic Research Agenda of the European Technology Platform Food for Life. FoodDrinkEurope, Brussels.

CIAA (2014). Strategic Research Agenda of the European Technology Platform Food for Life. Brussels: FoodDrinkEurope.

DBV (2011). Situationsbericht 2011/2012—Trends und Fakten zur Landwirtschaft. Accessed 8 October 2012, available at <http://www.situationsbericht.de>.

Eder, P. and I. Delgado (eds) (2006). Environmental Impact of Products (EIPRO)—Analysis of the life cycle environmental impacts related to the final consumption of the EU-25. Report 22284 EN to the European Commission, Brussels.

FAO (2009). The State of Food and Agriculture 2009 —Livestock in the balance. Rome: Food and Agriculture Organization.

FAO (2011). Global Food Losses and Food Waste. Rome: Food and Agriculture Organization.

Fritzen, S., S. Jarzebowski, and G. Schiefer (2012). Transparency on the Distribution of Value Creation from Selected Food Attributes in Relation to Cost Contributions and Margins at Food Chain Levels for Selected Food Chain Cases. Report D5.2 of the EU project TRANSFOP to the European Commission. Brussels: European Commission.

Fritzen, S. and G. Schiefer (2013). Feasibility of Transparent Food Pricing Distribution Mechanisms with SMEs. Report D5.3 of the EU project TRANSFOP to the European Commission. Brussels: European Commission.

IPCC (2007). Climate Change 2007: Synthesis report. Geneva, Switzerland: IPCC.

Nestlé (2011). So is(s)t Deutschland. Consolidated study. Accessed 2 May 2013, available at <http://www.nestle.de/Unternehmen/Nestle-Studie/Nestle-Studie-2011/Documents/Nestle_Studie_2011_Zusammenfassung.pdf>.

Parfitt, J., M. Barthel, and S. Macnaughton (2010). Food Waste within Food Supply Chains: Quantification and potential for change to 2050. *Philosophical Transactions of the Royal Society*, London B Biol. Sci. 27: pp. 3065–81.

Rabbinge, R. and A. Linnemann (eds) (2009). ESF/COST Forward Look on European Food Systems in a Changing World. ESF/Cost, Brussels.

SCAR (2009). European Commission (2009). New Challenges for Agricultural Research: Climate change, food security, rural development, agricultural knowledge system. EU RTD's Standing Committee on Agricultural Research, 2nd SCAR Foresight Exercise. Report to the European Commission. Brussels: European Commission.

SCAR (2011). Sustainable Food Consumption and Production in a Resource-constrained World. EU RTD's Standing Committee on Agricultural Research, Foresight report no. 3 to the European Commission, Brussels.

Schiefer, G. and J. Deiters (eds) (2013a). Mapping Formal Networks and Identifying their Role for Innovation in EU Food SMEs: A collection of case studies analyzed within the Netgrow EU (FP7) project. University of Bonn-ILB.

Schiefer, G. and J. Deiters (eds) (2013b). *Transparency for Sustainability in Food Chains: Challenges and Research Needs*. Amsterdam: Elsevier.

UN (2006). The United Nations World Water Development Report 2: Water—a shared responsibility. New York: United Nations.

9

Summing Up: New Insights and the Emerging Policy and Research Agenda for Addressing Food Price Inflation

Steve McCorriston

The contributions to this volume have aimed at providing new insights into retail price dynamics and the characteristics of food price adjustment across the EU by drawing on the recent experience of food price inflation across the EU, highlighting the potential role of the food supply chain and the nature of competition and other dimensions of the food chain in influencing food price dynamics and utilizing new data sources (i.e., scanner data sets) to provide new insights. These issues are at the forefront of the policy debate on how the food sector functions and how transparency in the food sector can be improved. Recent concerns focus on how commodity price shocks emanating from world markets impact on EU Member States and are influenced by the characteristics of the food sector while the current research frontier highlights the potential of scanner data to uncover the mechanisms of price transmission and retail price dynamics more generally. Scanner data may also relate to uncovering aspects of vertical coordination between stages in the supply chain such as the use of vertical restraints, private labels, and the monitoring of contracts that link agents at different stages of the food sector.

The issues of domestic food price inflation and the dynamics of retail prices have come to the fore in the wake of the world commodity price spikes of 2007–08 and 2011. Despite the widespread attention given to events on world markets, the corresponding effect on retail food inflation has varied across the EU. Given the potential impact on consumers (particularly the least well-off) and the differential impact on domestic farmers relative to consumers (where at the upstream stages, price changes are greater and more frequent), greater understanding of how the characteristics of the food sector impact on price transmission is a priority for both policymakers and the research community.

At the policy end, this has been reflected in a number of ways. For example, following the first price spike in 2007–08, the EU Commission document entitled 'A Better Functioning Food Supply Chain in Europe' highlighted concerns in the functioning of the food sector relating to 'structural weaknesses' and 'pervasive inequalities in the bargaining power between contracting parties' which 'delay necessary adjustments' in prices, 'prolong market efficiencies', and 'can exacerbate price volatility in agricultural commodity markets' (EU Commission, 2009: 4), serving to highlight concerns about how the food sector that links agricultural markets to retail food prices actually functions. The Commission working paper by Bukeviciute et al. (2009) highlighted concerns about concentration in the food sector at the retail stage and other features of the food sector in the price transmission process. The European Central Bank report (2011) highlighted characteristics of the retail sector in determining price transmission, while the European Competition Network report (ECN, 2012)—reflecting concerns about competition throughout the food supply chain—provided an audit of EU Member States' anti-trust investigations that focused on activities in the food sector. It should be noted here that while the policy debate and the contributions to this volume have focused on the EU, these concerns are not EU-specific, with similar points raised about food inflation and the role of the food sector elsewhere, including the US and Australia and covering developing as well as developed countries. As such, the research reported here as it relates to concerns about competition and other characteristics of the food supply sector—and how it relates to the behaviour of retail prices—has implications beyond the EU.

The research reported in this volume relates to these concerns linking the dynamics of retail food prices and price adjustment with characteristics of the food sector and points the way to further research that will improve transparency on how the food industry functions. There are two broad directions for future research that would provide important insights: one relating to access to data, the other to understanding how the food sector functions. These are obviously related but, for the purposes of highlighting priorities, it is useful to treat them separately.

Take the issue of price data first: although the commodities traded on world markets are often referred to as relating to 'food prices', they are not the same as food products sold at the retail end of the food chain. These raw commodities are only one input into the final processed good and can often represent a relatively small share of the value of the final product. Data are accessible on world commodity prices, and a consumer price index related to retail food inflation—as it can be for food sub-groups—can also be readily accessed. But there are two important gaps. First, the retail price data are often an aggregate reported at monthly frequency and—as we have shown in this volume—these monthly aggregate data are unlikely to fully reflect price dynamics at the retail level. Retail chains often have thousands of products on sale at any one outlet, and within product groups there is a high degree of product differentiation. This means that the retail prices that are reported do not fully reflect the underlying dynamics of retail price behaviour in the retail chain and, by extension, may not truly reflect the price transmission process from upstream markets through to retail.

Several contributions to this volume have highlighted the potential value of scanner data. These data relate to the specific product the consumer actually buys and are available at high frequency, by retail chain, by outlet, and across geography—though these characteristics are not always common to every scanner data source. Investigation of scanner data gives a different impression of the dynamics of retail food prices that are not available from the consumer price indices commonly available, and therefore has the potential to provide more detailed insights into price setting and price adjustment at the retail stage. But using scanner data also poses challenges (not least the cost of purchasing scanner data) and may also be limited to the questions that can be addressed if detailed price data are not accompanied by corresponding sales data. But the potential pay-offs are high in providing a more detailed account of price transmission by measuring product-specific elasticities (see Chapter 4), price transmission across space (Chapter 5), and across outlet, and in its relevance for measuring food price inflation (Chapter 6).

The second data gap relates to prices at intermediate stages of the food chain. The food sector represents a complex vertical chain from upstream raw commodity markets (world markets and domestic farmers), through food manufacturing and distribution, to retail. Studies that address food chain issues and the price transmission process are often restricted only to prices at either end of this food chain, with little or no information on price behaviour at the intermediate stages. As such, the detailed process of price transmission is a 'black box' and access to price data at intermediate stages would provide a significant step up in promoting transparency on how the food chain functions and detail more accurately the process of price transmission. Although EU Member States, in their price monitoring activities, are making some progress in observing and reporting these data, accessing these data for an appropriate time period and frequency would make possible more accurate accounts of how food prices adjust, and of who takes the burden of this price adjustment and why.

These issues of retail price dynamics and price adjustment through the supply chain also tie in with understanding the functioning of the supply chain, particularly the role of competition. But addressing competition in the food sector is complex. Given the vertical nature of food supply chains, and where different stages of the food chain may be characterized as oligopoly (such that we have a chain of successively oligopolistic stages), this implies that there will be both horizontal and vertical dimensions to competition in the food sector. As such, competition in the food sector does not solely relate to seller power, but also to buyer power and how the two interact. Moreover, this issue is not just limited to firm numbers (as reflected in concentration ratios): for example, the increased penetration of private labels may impact on the intensity of competition at the retail stage (the horizontal effect) but also impact on how manufacturers of nationally branded products interact with retailers (the vertical effect). Moreover, the dimensions of how firms at any stage of the food sector compete are also varied (for example, through alternative vertical contracts), while other structural changes to the food sector such as the increasing role of discounters, the penetration of private labels, the role of buyer groups, and so on, also impact on how the food sector functions and, in turn, on retail price dynamics and price adjustment.

In addition, concerns over 'fairness' are also prevalent, with the potential for bargaining power to impact on the weaker participants in the food chain, particularly farmers and small enterprises. More insights into the use and effects of contracts would also be relevant in addressing the functioning of the food chain. While bargaining power may influence the distribution of rents throughout the supply chain, there are concerns that contracts may also improve vertical coordination and, in doing so, circumvent the standard problem in successively oligopolistic markets associated with double marginalization and improve monitoring that can address potential market failures in relations between participants at different stages in the food sector. As the contributions by Bonnet et al. and Swinnen and Vandeplas highlight, addressing the vertical dimension of competition in the food sector also has implications for retail price adjustment. As a final observation, attempts to address competition in the food sector should also address producer and consumer welfare (i.e., the pro- or anti-competitive effect should not be limited to addressing the effect on consumer prices) and should also address dynamic as well as static concerns.

Addressing the various dimensions of competition in the food sector and how it functions across the EU is made more challenging by the observation that there is considerable heterogeneity in the characteristics of the food supply sector across EU Member States. In some cases, concentration at retail and food manufacturing is high, while in others it is less of a concern. High penetration of private labels is a feature of some EU food markets but not of others, while discounters have considerable market share in some countries but play a minor role in others.

Two final comments: as noted at the outset of this volume, increased concerns about retail food price inflation, the functioning of the supply chain, and the process of price transmission was fuelled by the price spikes experienced on world commodity markets. As we have experienced over recent years, price spikes come and go and, at present, world commodity prices have fallen and domestic food inflation is currently low. But issues about competition in the food sector will persist since the commodity crises have highlighted the current inadequacies in understanding retail food prices and how the characteristics of the food chain determine the overall functioning of food markets. While these issues pose significant challenges for researchers, it is an area where policymakers and stakeholders need to have a better understanding in order to make more appropriate decisions. In other words, the potential impact of high-quality and meaningful research is considerable. In addition, although the contributions to this volume have emphasized the experience of EU Member States, the issues of price transmission, retail price dynamics, and competition in the food sector stretch beyond the EU and it is hoped that the research contributions reported here can offer appropriate insights to a wider community of researchers and policymakers in other countries who face similar issues.

References

Bukeviciute, L., A. Dierx, and F. Ilzkovitz (2009) 'The Functioning of the Food Supply Chain and Its Effect on Food Prices in the European Union' European Economy Occasional Papers 47. Brussels.

EU Commission (2009) 'A Better Functioning Food Supply Chain in Europe' Communication from the Commission to the European Parliament, the Council and the European Economic and Social Committee and the Committee of the Regions. COM ("))() 591 Final. Brussels.

European Central Bank (2011) 'Structural Features of the Distributive Trades and Their Impact on Prices in the Euro Area' European Central Bank, Occasional Paper Series, No. 128. Frankfurt: European Central Bank.

European Competition Network (2012) 'ECN Activities in the Food Sector: report on competition enforcement and market monitoring activities by European competition authorities in the food sector'. Brussels: European Commission.

Index

commodity chains
  bread and wheat in EU Member States, 44
consumer price index (CPI)
  as a measure of cost-of-living, 123, 127–9, 131, 142
contracts
  contracting costs, 150–4, 156, 157, 160, 162–3
  contract-specific investment, 150, 163
  imperfect contract enforcement, 150, 158–62
  perfect contract enforcement, 155–8
  price transmission, 15, 37, 42, 67, 71, 75, 78–80, 89, 91, 92, 148, 150–163
CPI, *see* consumer price index (CPI)

dairy markets
  dairy desserts market (France), 15, 68, 69, 71–5, 77, 85, 86, 90–4
  fluid milk market (France), 15, 68, 73–5, 95
  Italian dairy market: price indices, 136–42, 144

econometric models
  panel estimation, 111
  probit quasi-fixed effects, 115
  random coefficients logit model, 75–7, 82–3, 91
  structural models, 15, 67
  time series, 31, 34, 38, 40, 54
  Tobit models, 53, 54, 56–7, 59, 62
  Tobit quasi-fixed effects, 115
  vector error correction models (VECM), 54
exchange rates, 2, 8, 21, 31, 43, 44, 48
expenditure share on food, 7, 123

food sector
  barriers to competition, 45, 46
  buyer groups, 9, 190
  buyer power, 9, 32, 36, 43, 190
  competition, 9, 10, 14–18, 21, 22, 24, 26–7, 30–9, 41, 43–6, 48, 67, 74, 78, 84, 92, 122, 140, 143, 148, 187–91
  consumer loyalty, 107
  contracting, 148, 187, 188, 190

discounters, 9, 44, 46, 47, 71, 73, 81, 90, 104, 190
double marginalization, 37, 42, 79, 190
food manufacturing/processing, 2, 9, 13, 15, 32, 37, 38, 42, 67, 175, 189, 190
food retailing, 2, 4, 6, 9–15, 21, 26, 27, 35, 41, 43, 44, 46–8, 102, 104, 107, 187–91
frequency of price promotions, 104, 110, 118
market structure, 14, 31, 36, 43
national brands, 15, 42, 66, 125
price–cost margins, 67, 75, 79, 80, 85–7
private labels, 9, 15, 42–4, 66, 90, 125, 187, 190
promotional strategies, 122, 137, 141, 143
rent distribution, 17
retail chains, 73, 75, 81, 123–5, 133, 189
retail formats/outlets
  hypermarkets, 102
  supermarkets, 2, 102
spatial competition, 122, 143
successive oligopoly, 36, 37
vertical restraints, 16, 41–3, 75, 187

German beer market, 16, 102–20

High Level Forum for a Better Functioning of the Food Supply Chain, 21

inflation
  'core' inflation, 6, 7
  food price inflation, 4, 6, 7, 10, 11, 20–2, 27, 40, 47, 65, 123, 187–91
  inflationary expectations, 6
  monetary authorities, 6, 7, 24, 28
  non-food inflation, 4–7, 14, 21, 22, 27–30, 48
  persistence, 7, 11
  variability, 28–30

measurement of inflation
  Fisher Ideal Index, 129
  Gini-Eltetö-Kövecs-Szulc (GEKS) index, 133, 135–7, 139–44
  Laspeyres index, 128, 129, 134, 136, 138–42, 144

Index

measurement of inflation (*cont.*)
   Paasche index, 128, 129, 134, 136, 139, 140, 142, 144
   scanner data, 122–45, 187, 189

pass-through, 31–7, 39–42, 46, 67, 68, 89–92, 156 *see also* price transmission
   cost pass-through, 42, 67, 68, 81–92, 102
   decomposing pass-through, 39–40
   long-run pass-through, 40, 67
price transmission, 2, 4, 9–12, 14–17, 22, 24, 26, 27, 29–43, 45, 47, 48, 51–63, 65–99, 147–64, 187–9, 191 *see also* Pass-through
   arm's length pricing, 37, 42
   asymmetric price transmission, 31, 40, 52, 92, 149
   branded products, 190
   buyer power, 36
   competition, 9, 14–17, 22, 26, 30, 31, 34, 36, 38, 39, 61, 67, 74, 78, 84, 92, 163
   contracting, 15, 37, 42, 67, 71, 75, 78–80, 85, 86, 89, 91, 92, 148, 150–60, 162–3
   curvature of the demand function, 40, 41, 67, 92
   determinants, 15–16, 66, 78
   economies of scale, 149, 150
   empirical studies, 22, 30, 51, 52
   Gardner model, 33
   horizontal price transmission, 4
   markups (and changes in), 10, 15–16, 32–4, 36, 40, 42, 67
   multi-product retailers, 41, 48
   non-linear adjustment, 31, 59
   non-linear pricing, 31, 37, 42, 79
   'over-shifting', 15, 42, 68, 89, 92
   price transmission elasticity, 33–6, 44, 45
   private labels, 15, 66, 90
   retail-farm level margin, 25
   spatial price transmission, 148
   speed of adjustment, 52–3, 57, 59
   successive oligopoly, 37
   technology (fixed and variable proportions), 32, 33
   'under-shifting', 15, 33, 68, 89, 92
   vertical price transmission, 4, 48, 51, 52, 55, 62, 148, 149
   vertical relations, 67–8, 75, 78–92
   vertical restraints, 16, 42

retail prices
   brand loyalty, 16, 108, 109
   dynamics, 2, 4–5, 11, 17, 21, 30, 187, 189–91
   frequency of promotions, 103
   promotions (sales), 103, 108, 115, 117
   resale price maintenance, 42, 79
   retail chains, 12, 16, 73, 75, 81, 102, 111, 123–5, 132, 133, 172, 189
   retail outlets, 2, 14, 16, 112, 178–80
   spatial pricing, 107–8, 111, 117, 119, 148
   temporal pricing, 16, 107

scanner data
   Kantar Worldpanel, 71
   measurement of inflation, 122–45, 187, 189
   Symphony IRI data set, 124, 126, 127, 137–45
specialization
   export, 57, 61, 62
   product, 15, 57, 59, 61, 62
supply chain methodology
   case studies, 167, 169
   networking, 167

TRANSFOP, *see* Transparency of food prices (TRANSFOP)
Transparency of food prices (TRANSFOP), 14, 18, 37, 57

world commodity prices
   cost (price) shocks, 1, 9, 43, 45, 46, 48, 148, 187
   oil prices, 30, 44
   price spikes 2007–2008, 2011, 1, 5, 20, 21, 187, 188